Engineering and Society: Working Towards Social Justice Part III: Engineering: Windows on Society

Engineering and Society: Working Towards Social Justice
Part III: Engineering: Windows on Society

Caroline Baillie and George Catalano

ISBN: 978-3-031-79954-9   paperback
ISBN: 978-3-031-79955-6   ebook

DOI 10.1007/978-3-031-79955-6

A Publication in the Springer series
SYNTHESIS LECTURES ON ENGINEERS, TECHNOLOGY AND SOCIETY

Lecture #10
Series Editor: Caroline Baillie, *University of Western Australia*

Series ISSN
Synthesis Lectures on Engineers, Technology and Society
Print 1933-3633   Electronic 1933-3461

Drawings © 2009 by Z*qhygeom

Synthesis Lectures on Engineers, Technology and Society

Editor
Caroline Baillie, *University of Western Australia*

Engineering and Society: Working Towards Social Justice
Part III: Engineering: Windows on Society
Caroline Baillie and George Catalano
2009

Engineering and Society: Working Towards Social Justice
Part II: Engineering: Decisions in the 21st Century
George Catalano
2009

Engineering and Society: Working Towards Social Justice
Part I: Engineering and Society
Caroline Baillie
2009

Engineering: Women and Leadership
Corri Zoli, Shobha Bhatia, Valerie Davidson, Kelly Rusch
2008

Bridging the Gap Between Engineering and the Global World:
A Case Study of the Coconut (Coir) Fiber Industry in Kerala, India
Shobha K. Bhatia, Jennifer L. Smith
2008

Engineering and Social Justice
Donna Riley
2008

Engineering, Poverty, and the Earth
George D. Catalano
2007

Engineers within a Local and Global Society
Caroline Baillie
2006

Engineering and Society: Working Towards Social Justice Part III: Engineering: Windows on Society

Caroline Baillie
University of Western Australia

George Catalano
State University of New York at Binghamton

SYNTHESIS LECTURES ON ENGINEERS, TECHNOLOGY AND SOCIETY #10

ABSTRACT

Engineers work in an increasingly complex entanglement of ideas, people, cultures, technology, systems and environments. Today, decisions made by engineers often have serious implications for not only their clients but for society as a whole and the natural world. Such decisions may potentially influence cultures, ways of living, as well as alter ecosystems which are in delicate balance. In order to make appropriate decisions and to co-create ideas and innovations within and among the complex networks of communities which currently exist and are be shaped by our decisions, we need to regain our place as professionals, to realise the significance of our work and to take responsibility in a much deeper sense. Engineers must develop the 'ability to respond' to emerging needs of all people, across all cultures. To do this requires insights and knowledge which are at present largely within the domain of the social and political sciences but which needs to be shared with our students in ways which are meaningful and relevant to engineering. This book attempts to do just that. In Part 1 Baillie introduces ideas associated with the ways in which engineers relate to the communities in which they work. Drawing on scholarship from science and technology studies, globalisation and development studies, as well as work in science communication and dialogue, this introductory text sets the scene for an engineering community which engages with the public. In Part 2 Catalano frames the thinking processes necessary to create ethical and just decisions in engineering, to understand the implications of our current decision making processes and think about ways in which we might adapt these to become more socially just in the future. In Part 3 Baillie and Catalano have provided case studies of everyday issues such as water, garbage and alarm clocks, to help us consider how we might see through the lenses of our new knowledge from Parts 1 and 2 and apply this to our every day existence as engineers.

KEYWORDS

engineering and society, social justice, ethics, engineering education, globalisation, public dialogue with engineering, engineering development, engineering studies

Contents

CHAPTER 1

Introduction

This volume is intended to be read in conjunction with the first two volumes of the series, 'Engineers and Society' and 'Making Decisions in the 21st Century.' We present a series of every day case studies, which exemplify the issues these volumes raise and we look at these cases through different lenses. It is possible to read these cases without first having studied the previous volumes although it is recommended that they be consulted at some point. In the first volume, it was explained that in these texts we take what is known as a critical perspective. In case you have not read the first volume, we will repeat some of the key features of this way of presenting material. If you have read Volume 1, skip the rest of the introduction and move straight to the cases.

Taking a critical perspective does not mean that we will criticise – but which we will critique through different lenses. 'Critical Theory' is an amalgam of philosophical and social scientific techniques providing a way to systematically, critically and yet constructively analyse systems of thought and practice. It includes a variety of possible approaches and perspectives by which to analyse 'not only cultural artefacts but also their contexts – social, political, historical, gender, ethnic' (Sim and Van Loon, 2000, p165). Critical theory is a tool which enables us to put engineering 'under the microscope' and to see whether we like what we see or whether we want to change it. For example, we might ask you to imagine that a large engineering firm is 'downsizing.' The management of the company need to consider the implications of what they are doing in order to make decisions. Workers are often faced with this situation and remain for many months or years unemployed after this event as they can find no work. This situation involves many complex arguments which we will explore later in the book. However, for now, to give an example of a critical perspective on this situation, we might explore the comparison between the individual versus social meaning of the situation. Osborne and Van Loon (1998) ask us to consider the question that some of these workers might ask, 'Why am I unemployed?' We might consider how much lack of respect an unemployed person has in society. Students of mine have actually commented that 'We can't help it if they don't respond to education.' It seems as if the common sense view is that the individual is responsible. In fact, there are many reasons for the surplus labour that causes people to be laid off which include:

Technological change (new machines).

Changed work practices (efficiency).

Work done in other countries (Globalisation).

Political change (government policy).

Cultural change (different products wanted).

Lack of requisite skills (no access to education or retraining).

None of these has anything to do with the individual worker but 'blaming the individual is common political practice' (Osborne and Van Loon, 1998, p10). Students will be asked to consider their own views and to debate and critique the various perspectives presented to them when focusing through 'windows' or lenses on typical areas well known to students. A very brief consideration of these windows will be presented below under the heading Lenses or Disciplinary Perspectives.

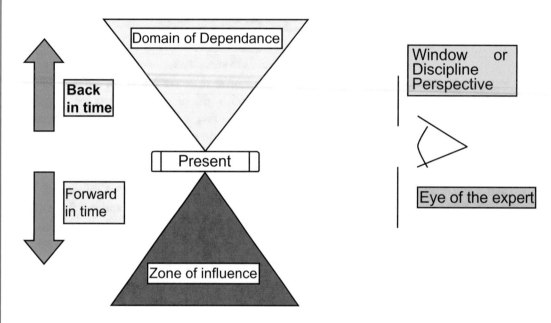

Figure 1.1. Windows of perception.

Perhaps then, we can extend the metaphor even further by placing various experts from different disciplines watching the unfolding of events through a window of a nearby building. Each expert will "see" the developing events within the context of a particular discipline or perspective. We will "see" that there are not events in history, sociology, political sciences, gender studies, economics, and global studies but simply events that can be understood in different ways and in different contexts. Each decision we make at this present moment in time is a result of the decisions and work of countless others at countless other earlier times. Equally as importantly, you will be introduced to the notion that each decision you make as a human being and as an engineer will have consequences that may extend far into the future and may be understood by others in a completely different way.

Let's reflect on the implications for our "zone of influence" and how we might relate it to our notion of "windows." The point we are trying to make is that what we see, what we come to regard as an objective truth, is a function of what we have experienced in our lives. An event which occurred during our childhood and how it effects our notion of reality is transported downstream much like the water that flows past the obstacle in its path. If this is true for events in our lives, then certainly, it makes sense to think of how our training and education can influence how we see the world, and this influence will continue to be with us as time advances. This idea can certainly be extended for the case of 'experts,' how they see the world and what implications these perceptions might have which are propagated forward in time.

Lenses or Disciplinary Perspectives

In the chapters that follow, we will not explicitly state each time we use a different lens, but to give you an idea of the different influences used to develop the thinking of this book, we present some of the typical lenses below.

Lessons from Sociology

This perspective frames how engineering is produced and utilized in local and global communities, how different engineering projects originate, how social needs are defined (participatory needs analysis), notions of social welfare and how the engineering impacts and affects different receptor societies. It also considers worker organization, employee relations, labour unions and social capital.

Lessons from History

This perspective deals with an exploration of the basics of the Industrial Revolution and the relationship between automation and labour. We consider the implications of this today with reference to globalisation and how it affects engineering organisations in developing countries. We will also explore the effect of increasing mechanisation on systems of thought and action.

Globalisation Debate

Current issues influenced by the global market will be considered using the work of key texts on globalisation - introducing the World Trade Organisation (WTO), 'free trade' global capitalism and sustainability.

Lesson from Economics

Different economic models will draw on a range of perspectives from capitalist, to Marxist viewpoints. Economic development will also be considered from the perspective of Amartya Sen's work on 'freedom.' His approach differentiates itself from traditional practical ethics and economic policy analysis, such as economic concentration on income and wealth (rather than on characteristics of human lives and substantive freedoms), utilitarian focus on mental satisfaction (rather than on creative discontent and constructive dissatisfaction) and the libertarian focus on procedures for liberty (with neglect of consequences of those procedures). There are, of course, connections to low income, but this needs to be integrated into a bigger picture. Poverty then can be seen as deprivation of basic capabilities rather than just low income, e.g., premature mortality, significant undernourishment, persistent morbidity and illiteracy.

Lessons from Political Studies

We will focus on the recent political changes that have occurred and the ways in which these affect the relationship between engineering and society. For example, the massive political-economic change that occurred with the collapse of state socialism, the establishment of private property, the influx of multinational capital into Eastern European countries and the integration of firms into the global economy.

Lessons from Environmental Studies

Recent developments on climate change, life cycle analysis, sustainable development, Kyoto protocol and different governmental perspectives will be explored and links to public health questioned.

Lessons from Psychology

From this perspective, we will consider personal development versus professional development. We will refer to recent work from psychology and Gestalt therapy, which helps the student consider their own blocks to self awareness, an understanding of personal ethics and potential ways of enhancing individual freedom and personal creative potential.

CHAPTER 2

Throwing Away Rubbish

In this chapter, we are focusing on the basic service of rubbish or garbage collection. As a citizen in the Global North, we expect our rubbish to be collected on a certain day of the week. As an engineer, we might reflect on the processes of recycling, transport, collection and separation. Viewing this process from the perspective of social justice, causes us to ask different questions.

Figure 2.1. Throwing away rubbish.

What happens when you throw away your rubbish (or your trash or garbage, depending on where you are from)? Maybe, if you live in a very environmentally conscious city, you will have separated your rubbish into those elements which can be recycled and those which cannot. You may even separate out organic or compost material. If you do this well, you will notice how little rubbish can be left that is destined for the landfill. For some people, however, rubbish is not such a simple issue. For many people it is a way of living. This chapter takes you on a journey through the maze of complexity that is waste in Buenos Aires, Argentina where we worked for six months during 2007. It is the story of a project, Waste for Life, coordinated by Caroline Baillie and Eric Feinblatt, and the context in which they found themselves. Along the way, they learnt more about waste than they ever thought possible.

2.1   THE POLITICS OF GARBAGE, BUENOS AIRES, ARGENTINA

After the economic crisis in 2000, 80,000 cartoneros or 'cardboard pickers' were found throughout Buenos Aires city, collecting up to 66 tpd of plastics, paper and other recyclables. A new law on 'Integral Management of Solid Urban Waste' which the group Greenpeace helped draft went, into effect late 2005. This law, better known as 'Zero Garbage Law,' was additionally intended to bring about decent working conditions for the many informal garbage collectors or 'cartoneros.' The law stipulates that the amount of garbage in landfills is to be reduced by 50%, and to reach that goal, the Buenos Aires city government has sponsored the organisation of cooperatives of garbage scavengers, and provided space for the first warehouse [World Bank Development (1999), Gobierno de la Ciudad (1998), Aiello and Grajales (2001)]. However, there are many other disorganised and family run cartoneros teams that still exist. Currently, the estimate is anywhere between 5,000 and 20,000. It is still the case that over 90% of the city's recyclables are collected by the cartoneros, informal and unpaid workers who live in the shantytowns and enter the city centre by night to find the waste (CEAMSE, 2007).

Much of the recycling that is collected is sold directly to agents at a price of about 16c/kg. Some more organised cooperatives sort and sell the materials directly to industry, and others sort and reprocess, and in some cases, recycle. However, few create final products, and therefore, income generated is very low. The potential is high for composites made from upgraded plastic, particularly forms which have found no other market, such as plastic bags and some plastic containers; however, no such technology is currently utilised in the recycling circles of Buenos Aires. Fibres in the city that could be used to reinforce the plastic include cardboard waste and wood chips.

Waste-for-Life is a collection of projects led by our team in Canada and focussed on supporting low income cooperatives to enhance their income by helping them to develop natural fibre composites from waste plastic, and locally found fibre using simple technology – a machine called a hotpress – which presses and heats the materials. However, these cost upwards of $50,000 to buy off the shelf, and we decided to design a lower cost version. Darko Matovic of Queen's University came up with the design before we left for Buenos Aires. Waste for Life ran in Buenos Aires between July and December 2007. We first conducted a series of needs analysis interviews and group discussions with local stakeholders such as cartoneros, local government, landfill managers and workers cooperatives. Was the idea of making composites from waste a useful one? Did the groups collect enough plastic? How would they wash it? Did they have storage space? Did they have a source of fibre? What product would they make, and who could they sell it to? Would it be cost effective for them? What equipment would they use to process the composites, and how could they afford this?

When we decided that we had enough interest from local groups; we commissioned a hotpress - the machine was built by a designer in the Buenos Aires area.....

Instead of going into more detail about the project, we have decided to show you an abridged version of the blog that was created during the time of the project. It is possible to follow the team's progress and learn with them about the complex political and social issues that we had to understand and cope with. We have deliberately kept the flow of the blog, so that you can read along and live with us in the streets of Buenos Aires, asking the questions we asked and seeing if you would have made different choices. It is a living example of all of the theory we have been speaking about in the first two volumes of this book. Many of the blog entries appear not to be directly about the project, but these are critical for our understanding of the environmental, social and economic context, and in our assessment, of needs. We have, additionally, superimposed challenge questions, after the fact, for you to consider what you would have done and what you might do in the future if you embark upon a project in a context very different from your own.

The Blog. (Unless otherwise stated posts and photographs are by Eric Feinblatt and Caroline Baillie, coordinators of Waste-for-Life, Buenos Aires, and written in the first person regardless of who was writing.)

2.2   FINDING OURSELVES

June 5, 2007.

Arriving in San Telmo, BA, after a very tiring and annoying journey complete with delays and lost luggage, we opened the doors to our new apartment full of character, shuttered windows, and the charm of new discoveries. The neighborhood, with its cobbled streets, market places, cafes and restaurants, was just what we had hoped for. After dark, we went on a search for some food and there, right in front of our apartment, were a group of 'cartoneros' – a woman stacking a cart with cardboard, a young boy ripping open bin-liners full of mixed and rotten waste to locate any morsel of paper or plastic. Why were we shocked to see this? That's what we had come for, wasn't it? But somehow we weren't ready. Some part of us wanted to hold onto that feeling of the tired, annoyed traveler wanting sustenance, warmth and comfort. The shock of our ridiculous, late (not even fully lost) luggage, one-meal-a-little-late, hungry problems embarrassed us.

We are more than aware that many foreigners think they can come and 'fix' problems in contexts other than their own. We do not wish to repeat the errors of others and hope for advice and partners who might help us with this project. The questions we thought about before we came keep reasserting themselves: Who should we look to partner with? What is the government's role in waste management and collection? Could our project possibly perpetuate a system of inequities? How will the proceeds from the materials and products the groups collect/make get distributed? Are we helping to launch an economic model that we ultimately do not support? How do we avoid profiteering from outside agencies as soon as there is a profit to be made? Who supports us? What affiliations do we have and how do they influence our values and our commitments? Is any of this our business anyway? Should we just stay at home?

Then there are the questions of the photographs and the website. What right do we have to upload photos of people whose life circumstances and choices we know little about? What is the purpose of our website anyway? To give ourselves a pat on the back that we are actually doing a project and progressing – even if progress is measured by images and chat? The cartoneros have a very public life as it is. They are on view every evening as we see them collecting their source of income. Furthermore, this is not just a Buenos Aires issue. Living off waste is a universal phenomena, complicated by governments who clean up only the nice parts of town and who recycle when it suits the economy. Before we left, we saw a scene that almost exactly replicated the scattered waste seen in San Telmo, traumatizing the pavements of our Bronx home.

Figure 2.2. Rubbish in Buenos Aires.

Figure 2.3. Rubbish in the Bronx, NY.

Challenge Box: What are your thoughts about what you have read so far? What right have we to come to Buenos Aires and think we can help? Locals can be quite upset when they think we are here because it is a city in need of development. Most portenos (people from BA) believe Argentina to be 'first' world.

2.3   KINGSTON HOT PRESS

Published by
Darko Matovic at July 17, 2007.

Here at Queen's University in Kingston, we are making the first press prototype these days (July 2007). The whole process is quite exciting for us all, but sometimes frustrating, of course. In the pages below, I will describe the press design, point out the things to watch out for while building the prototype, and describe the building process in a step-by-step fashion. As we gain experience testing and commissioning this prototype, I am sure that there will be changes necessary, but, of course, we are not aware of them now. As usual, every prototype is perfect on paper and on the screen, until built and tested.

The Kingston Hot Press (KHP) is a manually operated hot press designed to make tiles of composite plastic, up to 610×610 mm (24×24 in) size, 1-10 mm thick. The press is designed to provide up to 6 MPa pressure (870 psi). The total force required for pressing is thus around 200 tons (2 MN). The temperature can be adjusted up to 250°C. The key design challenge was to make a manually operated press able to supply very large force, and yet be affordable and relatively easy to make.

Here is what the current press design looks like:

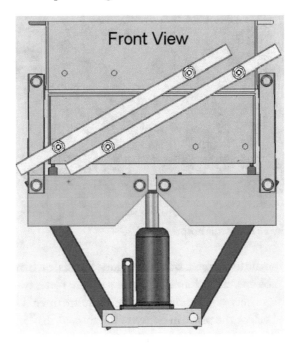

Figure 2.4. Design of the Kingston hotpress.

2.4   THE PRESS IN THE MAKING

Published by
Darko Matovic
July 18, 2007.

My current posts come from Kingston, Canada, where I teach at Queen's University. This is where the idea of designing and building the hot press came, after hearing about the efforts to startup composites production in (another Waste for Life) location, Lesotho, using local fiber and recycled plastic.

The hot press is essentially made of three functional units: (1) the lids, (2) the lifting mechanism, and (3) the stand. The "core press," consisting of the first two units is shown below in the triaxial view:

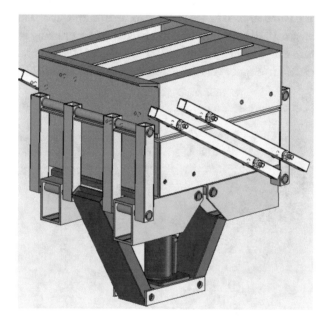

Figure 2.5. Alternative view of Kingston hotpress.

The top and the bottom lid are similar. Both are welded frames, each made of four "C" channels and two "I" (or rather "H") beams. Each frame supports a square plate, two plates pressing against each other, squeezing the tile in between. Inside the frame, there are three "U" shaped electric heaters in each lid, attached to the back of the plate, in between the frame "ribs," i.e., the steel profiles. The steel profiles are BIG, 200 mm high (8 in). This is what you need when trying to press a 0.6×0.6 m area with 200 tons, and not allow for more than 0.5 mm bulging in the middle. At least, this is what the finite element simulation says. We'll soon discover if this is right, not enough, or an overkill. This is what prototypes are for.

The top and bottom lid are similar but not identical. The top one has two C channels on the opposite side oriented outward, while the bottom has all four C channels pointing inward, mitred and welded as a picture frame. Here's the top frame in the making, photo taken today:

Figure 2.6. Making the press in Kingston.

2.5   WADING INTO THE WATERS

July 20, 2007.

We left North America two weeks ago with a handful of contacts in Buenos Aires, a sample 3inch x 3inch composite tile fabricated in Caroline's Queen's U. Lab, years of experience as educational developers, scientific knowledge about waste plastics and natural fiber composites, and a desire to share what we know with self-managed groups interested in developing poverty-reducing solutions to a specific ecological problem in BA. We presumed that our project would revolve around one or more of the cartonero collectives because they had access to the composite raw materials – the waste plastic bags and cardboard which are the material centerpiece of Waste-for-Life Buenos Aires – and because they are visible, often well-organized, but an at-risk population.

To the world outside of Argentina, El Ceibo, which is made up of 50+ families, is the most recognizable cartonero collective. It has been studied by university researchers and featured in the popular press, probably because of its tantalizing collaboration with residents of the up-scale Palermo neighborhood in BA. Cristina Lescano, it's founder, has more than 20 years experience as a successful community organizer, and we had been in touch with her through her 'secretary' Jim McAsey, a New York community organizer who had taken a year off to come live in Buenos Aires, inspired by the city's various self-managed collectives and, in particular, the 'recuperated' factory movement.

The waste plastic bags that the cartoneros collect have no value to them because they cannot be sold to recyclers. They end up in landfills or dumped in riverbeds on the outskirts of BA. Christine zeroed in on the commercial possibilities of 'upgrading' the plastic with cardboard fiber to create

Figure 2.7. The composite tile made from plastic and cardboard.

Figure 2.8. Cristina Lescano.

some sort of marketable building material. Of course, this would dramatically change the nature of El Ceibo's work from collectors and re-sellers of waste to manufacturers and distributors of finished goods. Such a focus shift raises all sorts of questions for El Ceibo and ourselves that we will be addressing in later posts.

2.6 INSPIRATION

July 22, 2007.

Last week, we also met with Dante Munoz and Carlos Levinton from the Faculty of Architecture at the University of Buenos Aires. Carlos directs the Centro Experimental de la Production (CEP) - Experimental Recycling Centre - and welcomed us as family members into its fold. We were introduced to the fabulous projects that they organize, including what they call social factories, working in poor areas, where they teach children new technologies so they can teach their parents. They have embraced all sorts of exciting recycling projects and educational experiments as well as the development of a recycling museum, now sadly closed. Dante and Carlos invited us to take part in a class for teenage school children one morning, and we really started to understand what their group was about. They are innovative, caring, socially minded, working with all manner of groups in different areas of Argentina and Bolivia. They function on the voluntary support of faculty and architects, as well as locals, who wish to get involved. It is truly inspiring to see this group in action. We were very moved by the commitment and love that is shared and that fuels this enterprise.

Figure 2.9. Carlos' group at work and one of his student workshops.

2.7 THE WORKERS' ECONOMY

July 23, 2007.

Over the last few days, we have been completely exhausted, stimulated, moved and frustrated by the conference 'The Workers' Economy: Self management and the Distribution of Wealth.' It is not news to the progressive that something interesting has been happening down here in Buenos Aires with the 'recovered factories' - those enterprises that were taken over by their workers when they went bankrupt in the economic crisis of 2001. The conference was hosted by the College of Philosophy and Literature at the University of Buenos Aires, and we were kindly invited by Marcelo Vieta, co-host and fellow Canadian academic originally from Argentina, to make a presentation about our project. We were now really entering very new territory, and it was with quite some trepidation that we entered the Spanish speaking world of commingled academics and workers from the recovered factories. Albeit wonderfully translated by a team of extremely patient experts, it was sometimes hard to concentrate on the endless 12 hour stream of papers that had been produced all on different facets of the same few intriguing factories. What made the meeting so memorable was, however, the commitment to the idea that here was history in the making, here was hope, here was imagination, here was a future that did not involve competition and selfishness but sharing and values which might be worth knowing. The workers themselves were the most profound when they spoke about their take-overs, their families, their problems and their dreams that they were really doing something worthwhile with their lives, that they could never have imagined possible. As we sat and listened to the representative from Hotel Bauen, the recovered hotel (which, despite its financial issues, was hosting those who could not pay during their stay in the city), a notice was posted that the Hotel workers were going to be evicted in a couple of weeks.

Figure 2.10. Bauen eviction notice.

All these hard working men and women, from all manner of organisations, told us that all they wanted to do was work. In order to break even, with all the accumulated debt of the owners, they were paying themselves pitifully small wages but were proud to be producing and serving. A highlight for me was the evening performance of theatre in the recovered factory of Chilavert printers.

Figure 2.11. Chilavert Theatre.

Community, culture and sharing is commonplace around these people. Students were flocking down to study this space, this phenomenon, and we were privileged to be among those who worked and studied in an area which seems to put flesh onto the starving bones of theory. On a personal level, we wanted to make contacts to see if any of the skilled workers in the factories might be keen to make our first hot press. We also met many academics who share common values and who we hope will help us understand the socio-economic-political implications of what we want to do.

Challenge Box: Consider the plight of the Recovered Workers. What are your thoughts about what they have done? How does this compare with workers in your own community?

2.8   THE PRESENCE OF ABSENT ONES

July 30, 2007.

Previously, we mentioned the friendly welcome we received from Carlos Levinton and his research group at the University of Buenos Aires. We'll write more about what promises to be a very fruitful collaboration with them in later posts, but want to point back to an unsettling experience for us when we first entered the School of Architecture two weeks ago.

Figure 2.12. Banner in the Faculty of Architecture of the 'Disappeared.'

Hanging above the lobby of the Faculty of Architecture is a stark black and white banner of young faces. They are pictures of students and faculty who were murdered and 'disappeared' during the dirty war of the late 70's and early 80's, and their faces are a grim reminder of the price of political opposition. Argentina lost a generation of university activists whose ranks are only now beginning to be filled by a cadre of incoming students, some of whom we were fortunate enough to meet shortly after our arrival here. We, in North America, have not experienced similar repressions or resistances, although we did have our Kent State, our police riot during the Democratic National Convention in Chicago, and mass university mobilizations against the Vietnam War. (I am aware that this sentence excludes the day-to-day experiences of many U.S. minority groups.) These have all but faded from our collective memories, and the notion of the university as a site of learning and engagement seems peculiarly anachronistic. In our talk at the Workers' Economy Conference, we spoke about the university and its original linkages to an educated citizenry who could maintain a democratic and just society. This is one of the ideas that fuels our work here in Argentina and that is continually being reinforced by the people we meet and the work that they are doing. It is an idea worth remembering and speaking about, as are the faces on the banner which simply says: 'siempre presentes,' always present.

Challenge Box: Find out more about what happened in the Military dictatorship in Argentina and just how many people disappeared for what they believed in. These questions as with understanding the recovered workers - help us to place ourselves in the context we are working.

2.9   LOST PLACES

July 30, 2007.

Ok, so I thought I knew how to use the 'Guia' – the incredible bus route map for the whole of the city. I was getting impatient and decided that we had to meet with some more cartoneros. We had contacts and I decided to call them up despite my horrible Spanish. After the call, I partially believed I had made an arrangement for 3pm and I thought I knew the address. The bus, however, had a different idea and decided to go west at a very crucial point. Another bus and a long walk later, we were in an extremely poor area, with private police seemingly patrolling a kids' football game. Large cameras and American/British accents were bad ideas. I was trying very hard to locate an address which appeared not to exist, amongst many stray dogs (and I didn't have that $600 rabies jab) and rubbish dumps. Finally, I asked a woman who didn't really know so I called the host and he explained to her and a growing throng of local kindly souls.

Several phone calls later we were in a small and very unreliable van driving over lots of holes to a totally different area of the 'favela.' We were told on the way that this whole area was supported by collecting rubbish. We were at the heart of the Dock Sur cartoneros patch. Our kind saviours deposited us at the right location, home of Carlos Perlini's Cooperativa de Trabajo Avellaneda Limitada. We were shown around the Galpon (warehouse) which included bottle stripping and weaving with a basic weaving machine. They also had equipment that could have, were they to have the funds to finish the machine, chop up plastic crates into small pieces which could be sold at a much higher price. Carlos told us they bought plastic waste from the local cartoneros and sold it - having washed it, dried it and packaged it. If they were to have the funds to support equipment they could create a greater value for the plastic they sold. This small, ridiculously under-funded facility supported 50 families.

2.10   SEPARATION AND SOLIDARITY

August 2, 2007.

Today, we went to the Bajo Flores recycling centre in another very poor barrio. The collective here has developed a separation plant - the first of four planned in Buenos Aires, where the members buy waste, delivered to them in trucks (which sometimes arrive at their destination), and separate it manually with the aid of a large conveyor belt system which enables them to separate plastic from glass and bottles from bags. The system is a standard design, which I have seen operating in the UK and Canada, and waste is further squashed into large cube shaped bales for selling on to the next in the chain. The equipment was apparently bought for the group by the city government, who also provided the warehouse. This is an interesting system. The city government, for the price of a warehouse and two pieces of equipment, gets much waste processed. No labour cost - as it is self run, except for a small subsidy to keep the peace. Is this the government shirking their responsibility? In the UK, U.S., and Canada, we expect the government to pay the salaries of our recycling workers. However, members of these self- run units are, seemingly, content to be their own bosses and fear that changes in the administration may mean they lose their livelihoods.

The most amazing thing for me is that people here seem to work together so well. We noticed this when we met the recovered workers. We were told that this was because they have become used to a lack of administrative support over so many years of loss and betrayal, that they have had to learn to support each other. I'm not sure if this can be the only reason, but we have felt privileged to work amongst these committed teams. Our host today, Silvia Rossi, another member of the University CEP, is, like many of the group, a self employed architect who works with the centre pro bono one day a week, and does many projects like this one in Bajo Flores on her own time. It was she who persuaded them to get the warehouse and equipment to get themselves started. She tirelessly supports them on a weekly basis. I will bring more than memories back from this project.

Challenge Box: Who do you think should be responsible for collecting and recycling rubbish? What do you think about the government using the cartoneros to do the recycling? In some cities this is even more well established such as Curitiba in Brazil.

2.11 WHO OWNS THE WASTE?

August 2, 2007.

The question of who owns the waste is a constant theme here in Buenos Aires. It is clear that the residents are made aware that they don't own it once its put in the street. The private/public organisations that run the trucks to collect the waste believe they own it, but their interest is not in helping the environment - they are paid to deliver to the landfill. Some of the workers break open the rubbish bags to relieve them of the recycling inside thereby adding many hours onto their roadside collections, but increasing their income by independently selling cardboard and bottles. The cartoneros who pick their way through the bags left on the pavements late at night, know they don't own the waste, but the tendency to make something from the discarded bodies and belongings has apparently been around for 200 years since the 'Ciruja' - when animals bones were made into useful objects. Some groups have evidently made arrangements that we don't fully understand, as they have access to the waste in whole districts without having to 'waste-pick.' Some are legal and some not, this is clear. But who does own the detritus, who do I give it to when I throw it away? I've given up the right, so who takes it? We have become so accustomed to idea that the government will take it away and that we pay them to do that through our taxes, so its a dirty and necessary job. We believe that putting our bottles in a separate bin means we are saving the planet, not realizing that much of this waste will be sold to a developing country that has no recycling system of its own. Their own waste then goes into landfill. The self employed, self organised cartoneros are doing the city's dirty work. They are amongst the few who are currently preventing the whole lot from going to the landfill. And they might earn a few pesos a day if they are lucky.

2.12 VARIATIONS

August 15, 2007.

Every time we leave the city we travel through the most incredible deprivation, the villas miseries or shantytowns. Houses are basic constructions, made of blocks and tin roofs, and many of these areas have no electricity or water and sewage services. 100,000 live off waste in the province of Buenos Aires, and many of the cartoneros travel from these areas into the centre of town to collect the city's detritus.

A visit to Maria Virginia of Abuela Naturaleza on Saturday proved to be extremely useful. She is an incredible 'entrepreneur' who has decided to spend her life working with waste recycling. She has been involved with this for 20 years, many years before the excitement about global warming. She teaches children in schools and fetes about recycling using puppets and other creative techniques and has at one time been involved with a cartoneros' cooperative. What she is doing right now, however, is providing a huge source of information and encouragement about recycling whilst making herself a living. She has organized her local community to separate products that can be recycled and goes around in a van to collect it. Although her van is at the moment being mended, she still receives visits from her neighbours with their deposits of recycling. Virginia has found a market for practically

Figure 2.13. Maria Virginia and her waste sorting depot.

everything she collects and showed us the huge number of items, all separated with the help of local workers that she employs. This is the most organized collection of recycled goods I have ever seen. Furthermore, she receives for most items about 1.5 pesos per kilo. Not much, but when you consider that the cartoneros receive about 20 cents per kilo for much of their produce, sold to agents who may sell it directly overseas, we learnt that here was someone who had found a way to actually recycle material and make a decent living. She hopes to pass on her knowledge to cartoneros' groups in the future. We also learnt about those materials that she cannot, at the moment, find a market for. These are a range of items made from plastic, cotton and cardboard, ideal for processing with our hotpress into composites. She even suggested a possible product for the materials.

2.13   BOTTLES ON CARS

August 15, 2007.

Figure 2.14. Bottles on cars.

We've been in BA for six weeks and during each of those weeks we've seen parked cars or pickups with large liquid-filled bottles on their hoods or roofs. At first, it was just a curiosity, and we had fun trying to guess what was going on, but eventually the persistence of the sight made it clear to us that this was a custom, not an anomaly, and that we didn't have a clue about the meaning of what we were looking at. This has been our experience with many things here.

Late the other night, we saw three CLIBA (Compania Latinoamerica de Ingenieria Basica Ambiental) garbage men rip open the large green plastic bags that they had thrown into their truck, sort the recyclables, and put them into their own 'for glass' and 'for can' bags - presumably to sell off privately. CLIBA is one of several private trucking companies that the city hires to pick up city waste and haul it to one of four CEAMSE landfills where it is sold by tonnage. CEAMSE (Coordinación Ecológica Area Metropolitana Sociedad del Estado) is the large public private organisation who owns all the landfills. This is a system that has many interconnecting players in it, each of whom claim a certain proprietary ownership of the garbage - for garbage, we are learning, is a very, very, big business here in BA. Four years ago, Mauricio Macri, the man who recently won the Mayoral election in BA, and who has a considerable commercial interest in garbage removal, accused the cartoneros of thievery because they were 'stealing his garbage.'

Figure 2.15. Garbage workers in BA (faces blanked out as they had been committing a 'crime' taking recycling from the garbage).

> **Challenge Box:** What do you know about where your rubbish and recycling goes to?

We spent part of a day last week at the Bajo Flores Green Point, the first of six projected model centers that the municipal government and foreign investors are funding around the city. Each of these centers is run by and for the benefit of a different local cooperative - often a cartonero collective - and is supplied waste to sort and recycle by one of the private garbage hauling companies. Bajo Flores is supposed to receive garbage from some of the city's five-Star Hotels and apartment buildings over 19 stories, but it was pretty obvious to us that the center was working way below capacity. What was happening to the garbage along the route? Though destined to Bajo Flores and the 40+ families who were working there, we learned later that the truckers or the hotel employees or the apartment building janitors were diverting it, and someone (who knows who?) was selling the recyclables on their own. There are lots of competitors for garbage and many people could profit making certain that Bajo Flores fails.

Figure 2.16. Workers at Bajo Flores.

Last week disappeared very quickly in a kaleidoscope of visits to all manner of extraordinary places outside the city. One visit was to La Vallol we where went to visit a 'Barrios de Pie' home building project. Except, there were no building materials and so no work was actually going on. Apparently, the money/materials from the government had not arrived, so empty shells of half-built houses lay dormant. Barrios de Pie is a politically astute and powerful association of workers' cooperatives across the country. Because they have an internal economy of sorts - one cooperative selling its goods or services to others - it is an appealing organisation for us to work with. There would be no need to identify or create an outside market for the composite products. For instance, the Barrios de Pie collective that picks waste plastic from the riverbeds, could clean it and sell it to the materials fabrication cooperative, which could process it into building tiles, which could be sold to the housing construction cooperative. And the hotpress that is needed to process the plastic could be manufactured by the machinists' cooperative. This would be a tight scenario where the beneficiaries of the composite product or products could be easily identifiable throughout the supply chain. Dante, who works closely with Barrios de Pie, took us to La Vallol, where the housing construction cooperative is working on building the first 16 of 180 projected homes. We wanted to look at the housing construction to see if there was any potential need for the composite tiles, but nothing was going on at the site - there was really no one to talk with. A few homes - partially finished - stood, and a few foundations were poured and ready for construction, but there were no workers around to work. The housing cooperative, which had won a contract from the government to build the homes, hadn't been paid the money needed to buy the materials, so work was at a standstill. Were we simply looking at a bureaucratic snafu?

The bottles? Oh yes, they mean that the vehicle is for sale.

2.14   THE PLOT THICKENS

August 16, 2007.

A meeting with the city government office, specifically the research group of the Politicas de Reciclado Urbano, proved quite a remarkable surprise. From what we had heard both before and after our arrival, we imagined a meeting with some dusty, defensive bureaucrats who would avoid all controversies and present their picture of success with academic splendour. Not a bit. Here was a group of four young researchers who really seemed to care about what they were doing - not just for the environment, but for the people. We can't say they represent the whole plan or the official view of the government, but they were a breath of fresh air. They were assigned to research the issues of waste in the city and told us much about the recycling centres, confirming the stories of the lack of success, non delivery of recycled materials and general confusion about where this waste does go. They also could tell us which reports had been written by whom and whether their studies on waste collection and recycling referred to both formal and informal sectors, domestic, street waste and other waste or just formal domestic. They estimated 11% of the city's waste was currently recycled, and that most of this was currently carried out by the city's 4000 cartoneros, along with El Ceibo as a major player. Five cooperatives have been singled out to run the recycling centres, as we had been told before, because they were the most organised. However, this was also the office, which registered the singles and families. So long as they were older than 15, we were told there were no other restrictions, this was not what we had been told before. They were given gloves and a safety strip to wear over their clothing, and so long as they did not disobey two basic rules, they could legally work in the streets. One rule was that when they ransacked the bags of rubbish they found there - that they would not leave the rubbish all over the street afterwards. And the other rule was that they would not stack up cardboard and plastic for their van so they could pick it up later. These two requirements are quite obviously not being obeyed by many groups as you see when you walk home at night, avoiding the spills of tomatoes that were strewn across the pavement. Furthermore, at many street corners on certain nights of the week, you would find a large pile of sorted recycling materials. Maria, Antonella, Mariela, and Felix were incredibly friendly, very concerned about the cartoneros and what might become of them after the forthcoming city government changes. It was difficult to doubt the sincerity of this group. We discovered that Mariela was a radio DJ for the Bajo Flores radio station. She told us she had to do something to feel alive and she went there as often as she could. The plot thickens every day.

> **Challenge Box:** Before we came to BA, we had been warned that it might prove difficult for us if we were seen to be affiliated with any group - government, private or social collective. So far, we had experienced a wary welcome by everyone, which turned into great friendship on only the second or third visit. Do you think it was the right thing to do - to remain unaffiliated? Why or why not?

2.15 SOFT COLLABORATION

August 17, 2007.

INTI (Instituto Nacional de Technologia Industrial), a federal government institution here in Buenos Aires, has some very intriguing functions. I would have imagined that it might be able to help us identify the location of a hotpress or advise us on other technical matters but, in fact, we stumbled across Hector Gonzalez at the Workers Economy conference and, finally, went to see him today. His job involves coordinating research, which supports the local cooperatives. Whatever comes in – they try to farm out to various of their departments. These are social, economic and technical issues. Instead of being greeted with a discouraging what-can-we-do-for-you smile – we were invited to meet others from architecture and construction. They had already downloaded our blog and were ready and willing to collaborate and help in any way they could. Our mission to help cartoneros seek extra income was also theirs, it seemed. We have now found quite a few left leaning government employees, all of whom are a little nervous about incoming city governments and the changes they will make. We were offered assistance to find the standards which will make our materials legal in buildings and in furniture, help in locating a hotpress or personnel to make one, the names of recovered factories who might make the press and finally offers of 'soft' collaboration.

2.16   NO COST HOUSING

September 3, 2007.

Shortly after arriving in Buenos Aires two months ago, we met up with Dante Munoz and Carlos Levinton. They have tutored us in the politics of waste and the re-purposing of waste (as well as ceviche and milongas), and we have met with them at least once a week in the Centro Experimental de la Production (CEP) at the University of Buenos Aires' School of Architecture, Design and Urbanism. Carlos is an architect, a professor of architecture and director of the CEP, which is supported by (mostly) volunteer architects and funded by a three-year, $7000/year grant from the University. The CEP has been described to us, and probably to University Administrators, as a research unit that investigates emergency disaster preparedness and response (like a FEMA think tank), but we are learning that it is much, much more than that and, in a conversation with Dante the other day, we began to get the bigger picture.

When the CEP thinks about minimizing risks to disaster, they are thinking about theoretical models and practical applications that reduce vulnerability to environmental, economic, social and natural disasters, and their work attempts to address some very fundamental questions such as: what is healthy housing? What are healthy cities? What are healthy exchange relationships? What impedes health, and what political, economic and educational activities are necessary to restore it? They are part of a larger, loosely joined movement of academics, collectives, and activists who, perhaps, inspired by the self-management movement that has flourished since the 2001 economic collapse of the Argentinean economy, or Argentina's history of legal and successful workers' cooperatives, or the scattered but concentrated experiments with Trueque (the Inca word for a multi-reciprocal exchange economy), believe that they can mobilize against the values of a consumer economy that processes identity and relationships in terms of acquisitiveness, accumulation and privilege. We share many things with the CEP, not the least of which are our mutual experiments in adding value to what is normally considered society's waste. We are working with low-density plastic (plastic bags) combined with natural fibers (cardboard) to create building materials and domestic products, and they are producing (in addition to many other things) solar heating panels and insulation from discarded 2-liter cola bottles. Both of us are investigating the use of low threshold/high impact technologies to improve peoples' physical living conditions and to reduce the proliferation of waste.

2.17   WATER TURNS TO MUD

September 3, 2007.

We have now been here for two months and I have the feeling that we are no longer wading into waters but though thick mud. I sometimes have no idea how to move or where to go next. I think this is an important feeling and I need to understand better what it means. Even speaking the same language in Canada (being British), I have the idea, sometimes, that I really don't 'get it.' Here with the language and so many different influences, historically, politically and economically, I realise how hard it is to work in another context.

The last week saw many meetings again, hopefully, leading somewhere but to where I am unclear. Also, we were joined by my cousin Karen's daughter, Emily, who is a geographer, and enormously helpful with the questions she is asking. We met with Brendan from Working World, who shared with us some fascinating insights about their incredible micro funding organisation and their views on the cooperative structures here. We also visited a very poor barrio in Moreno, where Carlos Levinton was teaching school children how to make hot water from waste plastic bottles - we even saw one of these 'machines' in action in a local house - the water was incredibly warm.

It was great to see how the work the CEP does manifests itself with the children teaching their parents. I had fun teaching some words of English to some of the children. They didn't understand why I spoke so strangely and when their teacher explained that I was English, they wanted a lesson. We exchanged words, naming bracelets and body parts and animals. I learnt as much as they, which was (to them) a surprise and much fun for all.

Figure 2.17. Learning English.

Our main focus this last week, however, was to visit two of the groups that we hope to work with. A return visit to Bajo Flores sorting unit, to discuss the next stages, and to make plans for a theatre group in their community centre (a sideline that Eric and I want to pursue here) and the most exciting visit - to UST (Worker's Solidarity Association) in Villa Dominico. We were hosted by our friend Marcelo and met Mario and his son who explained how UST works. It seems to be a 'recovered' enterprise in the sense that it extracted itself from the withdrawing corporation TechInt, and negotiated with CEAMSE to become contract workers, to maintain the regenerated land on top of a huge former landfill. They also work with seven other cooperatives amounting to 80 people who work in a variety of areas such as recycling and building. Furthermore, they told us have the

Figure 2.18. A solar water heater.

skills to make the machine. Mario was enthusiastic and immediately gave the hotpress plans to his colleagues to start thinking about the design.

2.18   TRANSFERRING OWNERSHIP

September 4, 2007.

Today we took the city office up on their offer to 'come up anytime the light is on' and asked more questions of Mariela, who is truly becoming our heroine. She knows so much and cares even more. I was interested to hear that cartoneros were commonly considered 'good' if they formed cooperatives, but that the chances of all cartoneros forming co-ops was extremely unlikely. We raced from there to visit Marcela de Luca, an engineering professor who works for both University of Buenos Aires and as a consultant funded by many different organisations including CEAMSE. We knew she had studied cartoneros and waste management in the city for years but had no idea what

Figure 2.19. UST Cooperative.

a wealth of information she would be. Again, she really cares about the cartoneros, but in a very different way to others we have spoken to. Marcela worries about the under 15 year olds who work alongside their parents and do not attend school. She gave us a figure of 9,000 registered cartoneros, higher than we had heard before. She also informed us that they had not really existed before 2001 - at least in their current status - different from the notion of ancient ceruja influences. Apparently, they were at work in the city area but not in the outer districts, and their functions were very different. They would collect directly from city offices and not so much from the streets. Marcela has created many, many reports and is extremely knowledgeable so we could only capture a tiny amount, but took away 6 tons of reports to read later (thank God for CDs). We also learned more about CEAMSE, which has become an enigma to us (even the city government say they don't actually understand how it works). It was started by the military government in 1978, seems to be a complex amalgam of city, provincial and national (maybe) government and private companies. They own the landfills, and as soon as the trucking companies drop off the rubbish that they have collected from the streets at transfer units, they own the rubbish. Until then, the municipality owns it. Are we starting to see the light? Not really. Who controls CEAMSE and who benefits and how privatisation works in this instance is intriguing to say the least.

Challenge Box: We have spent quite a bit of time trying to understand the context here in BA. Before I came, I was told by one well meaning social scientist 'stick to the science.' As you know (from Volume 1) there is no such thing as science in a vacuum. But how much did we need to know, and how did we know when we knew enough?

2.19   SORTING OUT THE DIFFERENCE

September 9, 2007.

A visit to CEAMSE's landfill, 'Norte 111' involved a tour around three sorting plants. The evolution of the tour is in itself a fascinating psychological journey. We got into a small air conditioned bus with some business men in white shirts and a very different mission and lived the experience of globalization within 30 minutes. The first visit was to the Social Factory on CEAMSE's property - run and staffed by cooperatives' members from the local barrio. CEAMSE gives them the waste, they separate it with rudimentary equipment and sell it, in order to pay themselves. One worker told us how lucky they were to have been selected to work in this place. The next location was a private factory adjacent to CEAMSE, which had more impressive equipment, and workers there told us that the compressor (to squash plastic and card into cubes to sell on) was the only one of its kind in South America. The final location was a Chinese run factory currently being established. We were shown into a huge warehouse, with machines resembling giant green dinosaurs waiting to be fed the waste that would come in tons. This was the largest scale I have ever seen. One of the businessmen suggested that the Chinese equipment was in fact old fashioned and would not work. He preferred a completely labourless, automated factory. Cartoneros watch out – your time is limited I fear.

A visit to CEAMSE does not necessarily include a visit to the landfill. As we were not official visitors, we had to use much persuasive power to get to the landfill itself. What CEAMSE did not want publicized is that the cartoneros still come in the hundreds every afternoon. They think that showing it on films and TV has increased the number of cartoneros that come in the hope of finding enough waste to live for the next day. At three thirty, they start to arrive - all ages – playing football while they wait, on their bicycles at the ready. When five thirty comes around, they are ready to race on their bikes the final two km to find the best of the recently dumped waste. Meantime, the animal-like machines wait in line to drop off their rubbish onto the open pit for the scavengers to explore…

2.20 REPROCESSING COOPERATIVES

September 9, 2007.

Two cooperatives funded by Working World, Etilplast and Villa Angelica, are examples of small family co-op organisations which buy waste plastic and reprocess it to add value to the chain. Villa Angelica has a yard full of waste and two machines, which chop up the plastic. When we visited, we were told that they had recently lost their only client and were hoping to mend their small extruder to seek new buyers. All the equipment in the yard was priced in terms of the number of tons of plastic waste that they had traded to get them. All these machines were worth less than a few thousand dollars. A more successful organisation, Etilplast, had an enormous extruder, which they had built themselves over some eight months. All machines there were huge, and according to Working World, a little over the top - industrial size machines rather than cottage industry. From our perspective, this group seemed as if it could make the hotpress, but how complicated they would make it and how long they would take, one could only guess.

Figure 2.20. Etilplast cooperative.

2.21 PRESSING ON

September 9, 2007.

Carlos Perini and his cooperative, Avellenada, was in a much healthier place this week when we visited. I wanted to make sure Carlos did not think we had forgotten him even though we still didn't have any firm plans for a hotpress, so we returned and met with several members of the co-op. In the last few weeks, they have stepped up their operation and have joined with another cooperative that has brought into the marriage a working plastics chopping machine. They can now collect, hand sort, chop, wash and dry plastic for sale by type. They had also discussed our project amongst all the members and were extremely keen to go ahead – assuring us that they have the skills to make the press and that all that is required is the funding to support the materials, the hotpress plans and training to make the composites. We have sent them the basic instructions to start things moving and wait avidly for the final plans of our press from Canada.

2.22   URBAN RECOVERERS

September 11, 2007.

Yesterday, Maria Virginia welcomed us again into her home in Ituzaingo. I am not sure why it surprises me that life moves on so quickly for our Urban Recoverer friends. Virginia had mended her van and was about to sell all her current supplies of recyclables, as she had to move things from her borrowed store (the next door neighbour's house) to her home. She clarified something that has been bothering me. Why was it that some of the cooperatives that we had met did not call themselves cartoneros when they clearly collected recycling materials? Virginia told us that the term 'cartoneros' really only refers to the very informal gathering that happens in the streets and, of course, to the most popular cardboard that they collect. The term Urban Recoverers is the official name preferred by those who collect from door to door and have a slightly more stable, organized existence. We had a very useful discussion to take our project on to the next stages and, like Carlos Perini, she is very keen to move ahead if possible. She thinks she has a contact who could build the press, she has a proposed product (recycling bins) and much waste which does not currently have a market.

2.23   BAJO FLORES

September 25, 2007.

Bajo Flores is one of six sorting centers or 'green points' that figure prominently in BA's Zero Garbage Law plans. We'll speak in detail about the significance of these centers in a subsequent post, but mention Bajo Flores here because it is where we have been running our weekly theatre workshops and because, in an earlier post, we noted that it was obvious to us that it was not working anywhere near its capacity. We went by yesterday for our normal 3 o'clock Monday rendezvous, and were told that it was cancelled because there was simply too much sorting being done. (Normally, the workers who participate in our theatre workshops are let off work early.) I walked into the sorting area and took some photos and videos of a very, very busy workplace.

What happened?

Bajo Flores was supposed to receive the raw garbage of BA's five-star hotels and some of its over-19-story apartment buildings. Two of the five private trucking companies and the single government-run company that are responsible for cleaning the streets and hauling the city's garbage were charged with bringing the waste to Bajo Flores to be sorted. The cooperative members separate the different types of plastic, the bottles, the carton, the paper, etc., from the other garbage - that is then hauled to the CEAMSE landfill - and sold to the industries that recycle it into paper and other plastic and glass products. In theory, it's a good system: the cooperative is responsible for sorting and selling, and it bypasses all of the middlemen that stood between its members and the final destination of the recyclables when they were working informally as cartoneros on the streets. The members share the profits equally; less garbage goes to the landfill, and the plastic and bottles and cardboard are reused. But it wasn't working, we were told, because the recyclables disappeared somewhere along the routes to Bajo Flores, and the cooperative (Bajo Flores Ecological Cooperative of Recyclers) was left with little to sort or sell. We learned, today, that the government decided to flex its muscles a little (it was their plan after all that was going awry) and called on their Secretaria de Inteligencia to get involved. They used GPS technology and planted some moles to track the truck routes and seem to have effectively disrupted the subterfuge. We'll see.

Figure 2.21. More activity at Bajo Flores.

2.24 SO MANY HOT PRESSES

September 27, 2007.

Whilst our partner Darko works away in Canada to try to get funding for future work (to build more hot presses in Africa), here in Buenos Aires, I try to understand his design and interpret it for the potential hot press manufacturers that I have been speaking to. Its not an easy task - I feel my materials' engineering background, together with my poor Spanish, leaves me disabled in the task of machine design. I have met some excellent mechanics and machinists over the last few days but as soon as I had the drawings in front of me, all the 'yes I can do that for sure' nods became, 'how does that work?,' 'I don't understand' and 'I can't do it unless I see the actual machine, the detailed instructions….' We hope, however, that we now have two potential builders for the machine and we press on with both. Today, I went to visit Victor who runs a huge workshop together with one partner and a handful of employees. He told me, he has been working as a metal worker for 14 years, and it quickly became evident how experienced he is. I was amazed by the size of the workshop and the extent of the equipment and capacity they have to create tools and machines of all kinds, but more so that they have built this from nothing in the last eight months. We spent a long time trying to understand the design, comparing it with the more usual four column varieties, which are extremely expensive.

Figure 2.22. Victor's warehouse.

We also compared it with the designs of an English model and a model developed by a former PhD student, Helen Cartledge and her brother, who has sent the designs to me by post from Australia. Victor was astonished at the amount of pressure we needed, and we had all sorts of fun deciding how the machine would actually work, but eventually, we got to a place where we think we can go ahead and start building. Tomas, our other hot press builder, is a designer who will work with machinists to realise the final product. He may come up with a different version. We have decided to go ahead with both - the race is on for the best hot press in town…In the meantime, I also visited INTI again today and finally met the Plastics group - and I saw my first working Buenos Aires' hot press. I am beginning to think that the only thing that has any real essence in this world is the hot press… Maybe I need a rest.

2.25   TRAVELING KNOWLEDGE

September 27, 2007.

We arrived in Buenos Aires almost 3 months ago (it's now the project's midpoint) with some fairly ambitious but, nevertheless, straightforward goals:

* Share knowledge of the processes of combining waste plastic with natural fibers to produce affordable composite building materials and domestic products.

* Use this knowledge to create opportunities for generating autonomous and sustainable revenue streams for some of Buenos Aires' most at-risk populations.

* Help reduce the damaging environmental impact of non-recycled plastic waste products.

* Create a 'how to' pictorial handbook (mass-produced and distributed royalty-free world-wide) that describes the methods, materials and manufacturing processes needed to combine locally collected waste with the fibers of naturally grown materials and transform them into useful building and domestic products.

We do have another, unstated goal, or perhaps it's just a question or a test or a self-indulgent exercise: Do two people, or three people, or whatever, have any chance of accomplishing 'development' (I use many quotes around this word) outcomes without the benefit of any pre-arranged alliances or financial or institutional support? We can't answer this yet, but we've certainly passed through some hopeful and less hopeful days.

Challenge Box:	Well, do they?

If you've followed our blog posts, you probably have an idea why we chose Buenos Aires for this project: the political framework of the 2005 Zero Garbage Law; Argentina's long history of workers' collectives and more recent history of self-management, and a 6,000,000 people strong multi-reciprocal exchange economy that thrived for a few years; BA's large, informal, work force of cartoneros who earn very little money but have access to the plastic and cardboard needed to make the composite materials.

Before we left for Argentina, our close friend, Jackie, acted as an intermediary to give us the message, 'be careful, the Mafia controls garbage in Buenos Aires.' We have heard some anecdotal information that this is true, but it hasn't affected our work, though sometimes, we spend a little downtime speculating on what this sentence actually means. We have, though, been very quickly sucked into the politics and economics of waste, and it's awfully difficult to get a manageable picture of what's really going on here. This is what we think we know, though we're sure the reality is much messier.

* The Zero Garbage Law went into effect in late 2005 with the intention of 'reducing as much as possible the garbage that goes to the landfills or is incinerated, to curb pollution of the soil, air and water,' said Greenpeace activist, Juan Carlos Villalonga. The law stipulates that the amount of garbage in landfills is to be reduced by 50 percent by 2012 and 75 percent by 2017 from 2003 levels. One of its ancillary benefits is that recyclable materials will no longer end up in landfills.

* Garbage collection in Buenos Aires is almost entirely privatized: the city government contracts 5 different trucking companies, each of which is assigned a specific district, and they are responsible for cleaning the streets of waste and hauling that waste to transit points or directly to the CEAMSE landfill. The single government-owned 'company' hauls waste from Buenos Aires' poorest district in the southwest of the city.

* Buenos Aires does not yet have a visible and/or official systematic recycling program. We've recently seen brightly colored CEAMSE plastic and glass recycling receptacles in non-residential areas of the city and have heard that there is a plan to put one of them on every city block. Residents would be responsible for dumping their recyclables into the containers, which would then be picked up by CEAMSE and....brought where, sold to whom?

* Buenos Aires generates about 4500 tons of garbage everyday. 11% of that garbage is disposed of by the 4000-20,000 cartoneros (we've heard both figures and many others in between), the city's informal garbage pickers; 97% of the city's recycling is done by those same cartoneros who make 80-300 dollars/month by selling the recyclables per kilo to intermediaries who then sell the material onto recycling companies.

* There are several recycling cooperatives (usually made up of former cartoneros) which buy, process and sell some of the recyclables, though the economics of these cooperatives is very, very fragile.

* The municipal government does have plans to build 5 more sorting centers or 'green points' scattered around the city. One is currently active. Each center will be run by a different cooperative and receive waste hauled to it by one or more of the trucking companies. They can also receive recyclables brought to them by cartoneros in trucks (only). The sorted waste is sold somewhere up the recycling food chain. Bajo Flores is one of these centers, and is run by a 50-member cooperative. 6 centers will probably not be able to support more than 300-400 cartoneros by giving them the opportunity to turn their informal work into formal work.

* CEAMSE (Coordinacion Ecológica Area Metropolitana Sociedad del Estado) has been in business for 30 years and is the biggest player in this whole story. It is a municipal and regional government amalgam with private affiliations, which manages the landfills that receive the waste of the 13,000,000 people of the greater metropolitan area of Buenos Aires. No one we've spoken to has been able to unravel either the structure or operations of CEAMSE, and opinions vary widely

about whether or not it is the devil incarnate. CEAMSE has only one landfill that is in operation, Norte III, a little over an hour's drive from the city center. (The others are either being turned into 'eco-parks' or have collapsed.)

We are quite insignificant compared to the major players here, which works to our benefit, and sometimes we lose focus trying to figure out what the 'big' story is. But we have met some small collectives who are collecting and selling plastic or washing plastic or processing it, and who make a meagre living from waste but are eager to step up to small manufacturing. With the help of our young intern, Nils, who is beginning a cost-benefit analysis; and with Victor and Tomas who have taken on the challenge of building 2 hot presses; and the folks at INTI who will allow Caroline to experiment using some of their equipment with, hopefully, The Working World, who can help put this whole package together for us; and Carlos Levinton and his group who will keep this thing going when we've left, our pieces are coming together and our story is sounding quite good today.

2.26   COOPERATIVA DE TRABAJO "19 DE DICIEMBRE"

October 10, 2007.

Yesterday, accompanied by Rhiannon, our amazing friend and translator, we visited the Cooperativa de trabajo "19 de diciembre." We learned about this 'recuperated' metallurgic factory shortly after arriving in BA, and it's name kept popping up over the past few months. After years of layoffs and a final shutdown, the bankrupt factory (originally named ISACO) was occupied by a group of its workers on December 19th, 2002 (thus its name), who formed a cooperative to recover the company and get it working again. It continues, as in its heyday, to manufacture auto components for the automobile industry and, ironically, one of its principle clients is it's former owner who has created another factory which buys and assembles the components and sells them on to the large automobile manufacturers (Ford, VW, Mercedes Benz). The '19 de diciembre' employs about 5 or 6 outside skilled workers who work the more complex die machinery and are paid $4/hour, which is twice as much as the 30 or so cooperative members earn. (It is typical for skilled workers to either leave or not join cooperatives because they can make more money as 'independent' employees.)

We are trying to prepare for what happens when we leave Argentina, and one of our thoughts was to see if any of the recuperated metallurgic factories would be interested in manufacturing and marketing the hotpress, which is the key piece of equipment needed to make the fiber-reinforced plastic composites. Two hotpresses are in production right now and, if they work (fingers crossed), our group will have successfully designed a piece of $50,000 machinery for under $1000! We think it might make economic (and solidarity) sense for one of the recuperated factories to manufacture and sell a hotpress at a small profit to any of the recycling cooperatives who want to use one, and at much larger profit to the plastics industry where hotpresses are commonly in use. We spoke about all of this with Enrique Iriarte, the Cooperative's president. Now, it's wait and see if the idea flies.

2.27 INTI

October 13, 2007.

We recently developed links with Patricia Eisenberg's group at the Plastics Department of INTI (Instituto Nacional de Tecnologia Industrial). Initially, we hoped to be able to use the hotpress at INTI to try out some of the local materials, but today, we met Patricia, and she was keen to move beyond this and to work in a much more collaborative way with our project. She mentioned that there needs to be sustainability beyond our visit to Buenos Aires, something we have been losing sleep over, so it was like a dream come true to hear her words. We are very excited that the INTI group might be able to provide training and/or technical support to cooperatives and possibly research new waste materials which might be used and for which there is no current market. Patricia proposed a meeting in the near future to develop collaboration points between Waste-for-Life, the team at UBA and INTI. We've scheduled one for next Tuesday.

Another exciting part of today's visit was finally having access to a hotpress and trying out the methods developed by my Queen's student Ryan Marien, who has been volunteering on this project from afar. Using his method, Adrian, Mariana and I were able to re-create the cardboard fibre composite process with local materials – including the-standing-on-it-to-keep-it-flat-whilst-cooling part!

Challenge Box: What are your thoughts on sustainability of development projects beyond their expected duration - i.e., end of term, end of design project, end of PhD, end of funding?

2.28 SYNCHRONICITY

October 13, 2007.

It seems to me that the conversations we are having now with cooperatives about our project are more detailed and to the point than they were when we arrived. Of course, you will say, because you now know more about what you are doing. But in fact, I haven't noticed that we present the project in any different manner, but more so that the cooperative members ask much more sophisticated and relevant questions now, as if it's the fourth time we have been to see them, rather than the first. It's as if the idea is in the air...Some of the details that have been discussed recently are absolutely key to the success of the work within any of the groups we hope to work with. Our intern, Nils, is creating a cost analysis template and many of the factors we have been researching for this were brought up in recent meetings, such as 'How much electricity would the press use?' 'What could we sell the product for?' 'How long will a ceiling tile last?' 'Who would buy it?' These conversations took place in our meetings today and yesterday in Cooperativa El Alamo and Reciclando Sueños, both very successful and active cooperatives working in the recycling business. Both groups have an incredible spirit and believe strongly that they are providing a service just as the garbage trucks are, so they see no reason why the trucking companies are paid huge sums of money and the cooperatives are paid nothing or small subsidies. Marcelo from Reciclando Sueños had a head on fight with the government about this at the recent conference we attended on recycling.

Reciclando Sueños is the first co-op we have seen which has actually moved on to manufacturing from collecting, sorting and selling. They have an actual product – a painting sponge, the handle and backbone of which is made from recycled plastic. They chop, clean, dry and injection mould the plastic, and then assemble the parts, selling them for 1.10 pesos each to a local wholesaler. We were extremely impressed with their team, who were supported by weekly workshops run by local PhD anthropology students. After a long cold wait in a draughty warehouse full of bags of plastic, we were privileged to experience one of these workshops. It seemed to us to be a cross between a recovered workers assembly (although we have only heard about these) and a cooperative development workshop. Members were asked to discuss certain issues and topics, including a series of case studies, which led onto some incredible discussions such as 'How should we distribute the benefits gained when someone is given an extra item by the neighbours such as a good pair of socks?' Sebastian and Maria, who have been working with this cooperative for over three years, facilitated the discussions brilliantly.

Cooperativa El Alamo is to become one of the city's Green points. We were shown around the future premises, currently damp and mosquito-ridden, and proudly told which room would be the sorting and which the compressing area. We were introduced to both groups by Gonzalo Roque of Avina foundation, who took us to El Alamo today.

2.29   RECICLANDO SUEÑOS

Published by Rhiannon Edwards October 13th, 2007.

On a rainy Thursday, October 11, I accompanied Caroline and Eric to the cooperative Reciclando Sueños located in La Matanza, the largest zone of the Province of Buenos Aires with about two million inhabitants. The cooperative, which has collected recyclable waste in the nearby middle class neighbourhood of Aldo Bonce since 2004, stood out in a big way from others that we have had contact with.

The cooperative is the epitome of do-it-yourselfism. It was very evident that the members of the cooperative had an extensive knowledge of the types of materials that they were working with, but this is knowledge attained through trial and error and collectivizing the knowledge learned from successes and failures. They've managed to build a few major and complicated machines to process

- cut into small bits, wash, and dry - the recycled plastic that they collect. This is important because the more capacity the cooperatives have to process recycled materials, the less they are at the mercy of intermediaries and the more value they can extract from the material. Apart from the impressive machines made from scratch and salvage to cut, wash, and dry plastics, they've also developed an injector machine that melts and molds a certain kind of plastic and produces the base and handle for a painting tool, which they will be selling in the hardware stores of the neighborhood where they collect recyclables. This is an important development from many angles. Economically, it serves the cooperative because they can get a much better price selling a finished product to retail stores than selling the raw material to industry. They also spoke about producing these tools as being important symbolically in the development of their relationship with the residents of Aldo Bonce, because they can show not only that they are providing an environmental service by ensuring that recyclable materials do not end up in the landfill, but they are using the materials collected from the residents to produce something useful that returns to the residents. The injector machine is essentially the same idea as Caroline and Eric's hot press in terms of the function it serves for the cooperative; it's just that it produces a different tool.

This cooperative stands out especially for me because of its position in relation with the government (the municipal government is the relevant organ) and its vision of the work that they are doing. Reciclando Sueños defines their work as a public service that the government has failed to provide. They see no sense in the current arrangement of the government paying large companies to collect waste. The service they provide is one that serves the environment and the neighborhood, especially with the return of useful products made from what they collect. They have refused, up to now, to accept (or rather apply for) any form of government welfare checks or work plans, insisting that their work is legitimate and they should be paid by the municipal government for the services they provide. This distinction is important, and it seems to me, important that the money be paid to the cooperative where they can decide democratically how to manage it - maintaining control over decisions of investment, wages, etc., - instead of what seems to happen elsewhere, that the government gives money to individuals, funneled through the cooperative. Reciclando Sueños approaches their relations with the residents, from whom they collect materials, from along the same logic, seeing their work as a service, and residents' participation as that of a party receiving a service that duly corresponds to them, instead of as an act of charity.

We had the incredible and unique opportunity to sit in on their weekly "workshop" where they discussed and resolved issues of the cooperative with some facilitation by the anthropologists, Sebastian Carenzo and María Inés Fernández Álvarez. Far from being preachy or manipulative as I, ever the cynic might expect from middle class academics, it was an hour of mutual respect and cooperativism in practice. They kept their facilitation to a minimum, mostly directing any input towards ensuring that the more soft-spoken cooperative members were listened to and clarifying, mainly through repetition and little to no content shifting, proposals and arguments of the members.

2.30   PAINTING A PICTURE

October 22, 2007.

Every Sunday night in the Bronx, before going to sleep, I take a few bags of separated garbage that have been accumulating during the week and put them down on the sidewalk in front of my apartment - plastic, glass, cardboard. These bags are picked up sometime between midnight and 7 o'clock Monday morning by city garbage trucks and …. And what? Well, frankly, I have no clue what happens to my bags of garbage and, like most people, just assume that New York City somehow takes care of it all. My responsibility ends at the separation and putting out phase, which is about the only connection I have with the disposal of what I consume, a process that is for the most part completely invisible to me. Sometimes, I see people sifting through the garbage looking for 'returnables' that they pile into their own bags or hijacked shopping carts and take to the nearest supermarket to get 5 cents on the bottle. I have no idea what this exchange represents in terms of their monthly economy, but I can't help but recognize that some people are interrupting my 'orderly' chain of consumption and disposal, bringing it to the surface, and making a few pennies off of it. If you multiply this scenario thousands upon thousands of times, you get the picture of what recycling looks like in Buenos Aires; only in that city, it is rare than anyone separates anything …. except the cartoneros.

The 2001 Argentinean economic crisis created vast, overnight poverty throughout the country and an entire class of citizens who lived off of garbage, which they gathered, consumed and, when they could, sold. It is estimated that 100,000 people still live off of garbage in Buenos Aires Province, and 4000 - 20,000 of them come in to the city proper to pick through the 4500 tons of garbage that is left on its streets everyday. Prior to the passage of the 'Zero Garbage Law' in late 2005, it was public policy to haul ALL garbage to the several CEAMSE landfills located on the outskirts of the city. (Not surprisingly, these landfills were/are located in the poor or poorest areas surrounding BA, and it is their residents who suffer the polluted water, noise and air that are part and parcel of the impact these landfills have on the local environments.) Nothing was formerly recycled; everything was dumped into the landfills. Not much has changed.

The purpose of the 'Zero Garbage Law' is to reduce the total amount of garbage going into the landfills, and since almost half of that volume is composed of paper, cardboard, plastic and glass, it is these recyclables that are really the targets of the law. On the surface, this looks like a simple environmental issue that can be solved over time with some public education and a few public policies, but because so many people are actually already making a living, as meagre as that living may be, from these recyclables, this purely environmental issue is really all mixed up with a very visible social issue as well. 97% of all the recycling that is done in Buenos Aires is being done, informally, by the cartoneros, and one question that was vigorously pursued at the recent "1er. Foro y Congreso Internacional de Políticas de Reciclado en Grandes Urbes" was why the government was paying 5 private trucking companies to keep the streets clean and haul all of BA's undifferentiated garbage to the landfills and not paying the cartoneros who were separating and often processing the garbage for recycling.

The cartoneros or urban recoverers who fan out over the city everyday often arrive by train from the outskirts (this is important because these special trains are, as I write, being shut down, effectively, depriving many of the cartoneros access to the waste that is their livelihood). These are the people who actually manage BA's recycling and, as best as I can tell, they can be divided into three groups - gathers, processors, and gathers and processors. The groups can be highly organized cooperatives, or tiny family 'businesses' that include parents and their children, or people working on their own, and they may have ties with the neighborhood from where they collect the recyclables, or work anonymously and have none at all. Some of the garbage is sorted and sold 'as is' to intermediaries and some of it is processed - either washed, or squashed or sliced and diced, and sold a little further up the food chain. But we're talking about pittances here, and at an average of about 500 pesos per month, most cartoneros are part of the 2.4 million people or 20% of GBA (Greater Buenos Aires) who live below the poverty line, which the government has pegged at 914 pesos per month. Our tiny intervention here could add a small, local manufacturing stage to the recycling process, which, if managed properly, would supply another source of income for the cooperatives.

2.31   PRESSING AHEAD

October 26, 2007.

There have been more than a few unsettling moments during the past 3 1/2 months. Most of them have to do with the recognition that we have round-trip tickets and will be leaving here, that we will be alright. But the people we work with and have grown so close to can't really go anywhere, and their futures are much, much less certain. Perhaps, the biggest drag on moving forward, the thing that has the potential to finally torpedo everything we've been doing here and turn Waste-for-Life into an academic exercise, has been the hotpress - or lack of a hotpress - which is the key mechanical technology we're sharing with the recycling cooperatives. Getting the hotpress manufactured here (as well as in Canada) has been a major stumbling block, but yesterday, we visited Tomas' workshop, which was a 3 hour trip for us out of the city, and saw the pieces of the hotpress ready to be assembled. We expect that we can actually begin training sessions within the month.

2.32   THE PHOENIX OF LANZONE

October 26, 2007.

Renacer Lanzone or the Pheonix of the barrio Lanzone is a social organisation run by Adam Guevara. Adam had an idea over twenty years ago, that CEAMSE should send the trucks to groups of cartoneros, before putting the garbage in the ground, so that they could take the recycling materials, and leave less waste to be buried. He tried with many different Presidents of CEAMSE but it was not until the present one agreed, over three years ago, that Adam was able to realise his dream and set up a social factory or sorting centre in Lanzone district. His project was so successful there are now four centres and four more planned on the CEAMSE site – some of which we had already seen. Adam's group are very successful, he told us, because they have been trained by INTI - the

Figure 2.23. Buenos Aires hotpress in the making.

same plastics group that we work with, to separate the plastic waste very well and to sell each type individually. They make a lot more money than the other groups as a result – up to 20 pesos an hour. Adam cares very much for his group, and when I asked what criteria he had for choosing people to work there, he told me it was on the basis of need – if they have more children, etc. He even keeps a section of the factory as a store of PET bottles to become a Christmas bonus at the end of the year.

Training seems to be an important part of the success here. INTI training the technical selection process and also Suarez, a local professor, and Neumann, a consultant engineer, working for the University of Sarmiento and CEAMSE, training them in roles and responsibilities, team work and conflict resolution, the context, history of recycling and cartoneros as well as the more technical matters. Adam is very keen to move on with us, and he seems to be an extremely innovative thinker. As wary as we are of CEAMSE, we would love to work with Adam's Pheonix.

Challenge Box: At this point, with six weeks to go, we had no idea whether the project would work, whether we would have to come back, whether INTI would support the work, if the hotpress would work, if the cartoneros would find a product to manufacture.... What would you do next?

2.33   THE MIGHTY HOTPRESS IS FINISHED (ALMOST)

November 17th, 2007.

We passed by Tomas Benasso's workshop a few days ago to gaze at the finished hotpress and go over some of the minor kinks that he's going to work out during the next 2 weeks. It's a mighty beast of a machine and a tangible measure of one of the things we're trying to accomplish here in BA. The very good news (besides Tomas' heroic work from Darko's design) is that INTI's Plastics Group has agreed to house the Kingston Hotpress and begin putting it through it's paces, which will include property analysis and testing samples against local building and product codes and standards.

Figure 2.24. The hotpress in Tomas' workshop.

2.34   SIN PATRÓN

December 12th, 2007.

While waiting for the final tweaks to the hotpress (by the way, Tomas has tweaked it and is delivering it to INTI this Thursday, December 13), the remarkable Erika Loritz has helped us conduct street interviews with cartoneros.

We have spent most of our time in BA working with cartonero recycling cooperatives. They have the structures in place to successfully incorporate a small manufacturing channel to their, mostly collection, sorting, processing and distribution operations. But equally important to us is a common political and social outlook that stresses equity and human interdependence, and we have been the privileged witnesses of these ideas in motion, particularly their commitment to wage equality, non-hierarchical participatory decision making practices, and to 'socializing' their knowledge. We did not come to Buenos Aires to be community organizers, and so the individual or small family groups - the informal cartoneros who do much of BA's recycling - have been, for the most part, a closed book to us. We realized, early on, that we were not hearing their voices, although we did hear many opinions about them from other people, but we didn't have either the language skills or the confidence to simply go up to these people on the street and start talking. I spoke about this with Erika in Iruya, and she immediately offered to help us take up this challenge.

We have done 4 interviews so far and have others to do, so we are not working with a large dataset; however, there is a strong leitmotiv that we can point to, that we've been aware of ever since attending The Workers' Economy: Self Management and the Distribution of Wealth conference in July, and that can be summed up with the phrase, 'sin patrón.' This phrase, which means 'without a boss' or 'without an overseer,' gained currency during the recovered factory movement that began with the 2001 economic crisis, but we heard it used over and over again by the cartoneros we interviewed to describe the essential quality of their lives in very positive terms. They did not want to work for anyone, they did not want to be subject to the collective decision-making processes of a collective, they did not want to bother anyone, and they did not want to be bothered. It is still much too early to sum up the content of the interviews, and we have a lot of transcribing and thinking to do, and loads to read up on about informal work, but with Erika's help we have a special opportunity to learn something about this previously unavailable population of urban recyclers.

2.35   CHAMPAGNE!

December 16th, 2007.

Tomas delivered the Kingston Hotpress to INTI on Thursday as scheduled, and after about 2 hours of fiddling around with the electrical installation, we had a go of it with the strips of plastic bag that Caroline had been cutting up all morning - a task that was as much a necessity as occupational therapy, seeing that we were all sitting around on the edges of our seats. The hotpress had never been road tested - it only worked, theoretically, according to Darko's calculations and ingenious SolidWorks design. This was a lean and mean machine and, in order to keep the costs down, had no gauges (gadgets) to let us know that it actually generated enough heat ($160°C$) and enough pressure to melt the plastic. From my non-scientific mind's perspective (after all, this was only the second time I had ever been in a research laboratory) the first test was a disaster. We were all much too impatient and didn't let the press heat up enough - we expected this would take about 20 minutes, but only gave it 10 - and didn't leave the plastic in the press long enough - we thought this would take about 15 minutes, but only gave it 7. We cooled the mold down with water and opened it to a soppy, un-melted plastic mess. Our faces were all frowns. At this point, Patricia Eisenberg leaned over to us and said, 'don't worry, we'll make it work,' and her INTI crew stepped in, inserted a funny little L-shaped thermometer into a gap in the press, declared that it was indeed hot enough to melt plastic, poured a bucket of polyethylene pellets over the few strips of plastic bag that remained in

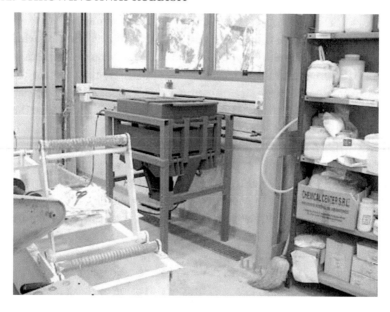

Figure 2.25. The hotpress in the lab at INTI.

the mold, spread the pellets out evenly (kind of) over the 60 cm × 60 cm surface, and shoved the mold back into the press. Tomas pumped up the car jack that created the pressure, and I was the timekeeper. 15 minutes later, Tomas released the jack's pressure, and we took the mold out of the press, laid it on the sidewalk just outside the building's side door, hosed it down, and opened the mold to reveal a slightly rough around the edges but beautifully hardened 2 kilo plastic plaque. The hotpress worked, just as Caroline had always believed, said, knew it would, and we've invited members from the 9 cooperatives we've worked with during the past 6 months to come by INTI on Wednesday to see it in action. Six months after arriving in BA, it's time for champagne!

2.36   HANDING IT OVER

December 19th, 2007.

In a few hours, we're off to INTI to participate in what is surely our final Waste-for-Life meeting before leaving Buenos Aires. We've already begun saying goodbye to our many 'compañeros,' a word that after 6 months of learning and struggling here has real meaning for us and is not at all embarrassing to use. And though this may be our last meeting, it is really a first because we are bringing together in one place all of the players who will keep Waste-for-Life BA alive after we return to North America - representatives of the 9 cooperatives we've worked with since July 2007; Carlos Levinton from the University of Buenos Aires' CEP; Gonzalo Roque from the Swiss NGO Avina, and Patricia Eisenberg's INTI plastics group. (The only person who can't make it today is

Figure 2.26. The first misshapen attempt at molding.

Esteban Magnani from the micro-credit organisation The Working World (La Base), who we met with yesterday and proposed a collaboration scenario which he enthusiastically endorsed.)

We've just returned from the meeting which, because of the collective engagement and emotion, caught us pretty much off guard. We had clearly handed over Waste-for-Life to the local stakeholders and could feel OK about going home. There was a single moment in that research laboratory, amidst the hotpress and all of the other grey testing equipment, and all those people who had such different stories to tell, but each of whom had been drastically affected in one way or another by the last decade of Argentinean history, when it became crystal clear why we were doing what we were doing. Adam Guevara talked about what he had learned from the INTI scientists and how it had changed the life, yes, the life, of the 20+ members of Renacer Lanzone, the civic association that he runs. These people collect and separate and sort plastic, which is a stinking job, and a couple of INTI scientists from Patricia's plastics group had spent time with them, some time ago, teaching the group how to do their job better by being more precise in the identification and, thus, the separation and sorting processes. Adam's group, whose members come from the shantytowns across the highway from the huge CEAMSE dump, took these lessons seriously, and now are able to sell their plastic for 4 times as much money as any other recycling group we've met. It was Adam's first opportunity to thank the INTI scientists, which he did with great dignity, and which sent some of them out of the room in embarrassed, unexpected tears.

Figure 2.27. The team at INTI - looking at the press.

2.37   WHO'S DOING WHAT?

December 19th, 2007.

Who's doing what? How will Waste-for-Life BA sustain itself? The only thing we knew for certain before coming to Argentina was that our stay was finite. The time that separated our flight into and out of BA was 6 months; we knew very little else. It's premature to map out what we've learned and done, and way too early to draw any conclusions, but one thing we can point to is that we've adhered to our idea of Waste-for-Life as a loosely joined network of people with diverse competencies, sharing common values, who join together to work through poverty-reducing solutions to particular ecological problems. We had no a prior idea what form this would take, but this is what the decentralized structure looks like to us as we prepare to leave.

Levinton's CEP is taking care of the experimentation in BsAs. They will spend the next few months putting the hotpress through its paces using different types of plastic waste in combination with cardboard and other natural fibers that they will collect from the cartonero cooperatives. They will test the results for compliance with local building codes and standards and will work alongside the cooperatives to teach them how to make the reinforced plastic composites. The project will be supported by our research team in Canada and once we have nailed down the technical processes, CEP will begin product development with their own hotpress, seeing what useful and/or fanciful products they can tease out of it. They will work closely with the cooperatives, receiving and testing out their ideas and helping their members with production. Avina is waiting in the shadows for one of the cooperatives they work with to propose a manufacturing and distribution plan that they can support with their funding and facilitation resources, and it is very likely that they will be behind the first complete test case. The Working World (La Base) will help identify and work with the cooperatives, which have the most likely chance of developing a successful manufacturing channel, and will establish and administer a revolving hotpress loan fund to enable the cooperatives to purchase a hotpress. And the cooperatives will be the innovators, the inspired ones, for they are the ones who have hope, who struggle every day, who work so hard with so very little yet can stand on their own shoulders and see beyond what they do now into a future where they deserve and will have a little bit more.

2.38   FINAL THOUGHTS

This has been rather an unusual chapter. It comes, as you can tell, directly from a blog so you have been able to track the movements of our team as we worked with the cartoneros in Buenos Aires. It is clearly an ongoing project. Many questions have been raised and many left unanswered for you to ponder. There are no right answers about any of these issues – but there are ways of thinking that are more helpful than others. Making sure that you think things through and question your prejudices and assumptions will always be a good thing to do. We had to do this constantly. Even as we left BA and sat in the airport wondering at what we had done, we had many elated thoughts such as 'How did we ever get to elicit the help of such amazing people?' 'How does Adam Guevara develop such a giving attitude when he has so little?...' We also had some troubling thoughts – 'Would the groups, together with the support of the University and INTI, actually be able to manufacture a product worth selling? Would they be able to sell it? Would we be able to get an organisation to make more presses? Would this make a difference to the cartoneros' lives as the knowledge about sorting plastics, that Adam's group had gained from INTI a few years before had done? Were we leaving too soon or was it time to go and hand over the tiller? None of this seemed to matter. We had handed the project over without even knowing it. We were never in control of this work. There were so many factors, so many features to think about, and we had tried to react to everything and think of as much as we could. We were delighted when Estaban from Working World said to us a few days before we left – 'you did it you know, many come here with great ideas and they just don't work.' But what had we done and how had we done it? We asked ourselves this question time and time again. The lessons learnt were many, but in the end, we decided the most important one of all was that we had listened; we had sought out as many different voices as we could and we tried to remain unbiased whilst staying within our non negotiables. These were: working with a participatory, inclusive approach, with the aim of poverty reduction, and the enhancement of social justice and environmental sustainability. We felt that all the various players who had learnt over time to trust us and to work with us had understood that this and nothing else, other than our own personal satisfaction with doing what we enjoyed, was all we had to gain from this project.

Challenge Box: Think about what the team achieved in the six months. What more could they have done? What could they have done differently? What should they do now to make sure that the next steps work?

CHAPTER 3

Turning on the Tap

Figure 3.1. Funnelling public water for private sale.

3.1 TURNING ON THE TAP

Its very easy to imagine, when we are living in England, the U.S., Canada or other parts of the world where water is in abundance, that everyone can just turn on the tap as we can and water will flow. However, we are starting to understand that water is not always available and that we need to be careful with how we use it. Australia has seen a drought for several years at the time of this writing and Australians are now shocked to see water being used to water gardens in other countries as this was banned a long time ago in their country. In England, a country, which seems to see rain all the time (although I suppose it actually only dribbles), we had a drought in the 70's which meant that we had to use pipes in the road as water was cut off to individual houses. In Book 1, Chapter 4, we saw what happens when droughts are badly managed and the devastating effects of famines come upon countries. 'Water scarcity may be the most underappreciated global environmental challenge of our time' (Barlow, 2003, p3), (World Watch Institute, 2006).

Figure 3.2. Children playing in fountains in Salt Lake City.

Young people who have not directly encountered droughts of any degree can easily ignore the sustainability of water as something not relevant to them. I once asked a class of students to write about creating a sustainable city - what would they need to consider and how would they do this. Despite the clue of the project topic, all 600 students, when I asked them to describe 'water-use' in their city, described how they would create marinas and water sports facilities. When I spoke to one of the university governors about this in passing, she said 'Oh yes, my children would be the same as they are so protected and lucky here in Canada.' The very important topic of sustainability is not one we will focus on in the section, although it will arise in the discussions. We are going to look in more detail at the issue of privatizing water. When I was young in England, we did not have to pay for water. Water was a free resource. It still is to me. I was shocked when we had to start paying for the amount if water we used. It just felt wrong. In the discussion below, we look in more detail at the differing views and implications of privatizing water in different parts of the world.

We will see that there are many different views on this topic, and water privatization has had a powerful effect on the way we think about ownership of natural resources. Some say we mine oil

Figure 3.3. Blue gold.

and gold and we sell that and we also sell trees, which grow on our land so why not sell water too. Others, the authors of this book included, feel that it should be a basic human right. Suffice to say that there are many people in the world without access to water, either because their village has not yet been developed enough to be connected by pipes and pumps to a water source, or because they cannot afford the pipes, or because they cannot afford to turn on the tap, and it has been cut off by the company who now 'owns' the water in their area.

> **Challenge Box:** Track your water consumption for one day and compare with your friends. If possible try to find out how this compares with someone from a country where water is less abundant. What differences are there?

3.2   BLUE GOLD

In Canada, Maude Barlow is something of a heroine, fighting for the rights to water. She has written some very useful texts, 'Blue Gold: The Fight to Stop the Corporate Theft of the World's Water' and 'Blue Covenant: The Global Water Crisis and the Coming Battle for the Right to Water' (Barlow, 2003, 2008). Her basic facts are helpful and continue to amaze us. 'The average human,' she tells us, 'needs fifty litres of water per day for drinking, cooking and sanitation. The average North American uses almost six hundred litres a day. The average inhabitant of Africa uses six litres per day' (Barlow, 2003, p5). But how can we help? Think most Canadians. We hear: 'Its not my fault that we have more water here.' But virtual water is a huge problem. This is water, which is displaced through the trade of goods which needs water to grow or manufacture them. Marlow tells us that Kenya is using up the water of Lake Naivasha to grow roses for Europe. Biofuels, for all their environmental promise, use up land which used to be used for crops, but also use massive amounts of water. It takes 1,700 litres of water to produce 1 litre of ethanol. 'Vietnam is destroying its water table to grow coffee for export' (Barlow, 2003, p17).

But what has this got to do with engineering? Sadly, everything – in many direct and indirect ways. The technology used to manufacture the biofuels was developed by well meaning scientists and engineers thinking that they were helping to reduce global warming but without fully understanding the implications of their creation. The huge increase in manufactured goods uses up water, which we don't even think about. Bottled water is one of the most polluting industries that exist. They have become a fashion accessory and healthy outdoor types love to carry a bottle around but the reality is that one million bottles of water cause emissions of 18,000 kg of carbon dioxide and fewer than 5% of bottles are recycled (Barlow, 2003, p99). I trained in materials' engineering, but I did not learn how much water was needed to create the plastics and metals, which we make everything from.

Even when strategies have been developed to try to alleviate the water crisis, Barlow informs us that the high technology solutions, the dams, diversions and desalination are part of the problem. Larger dams 'significantly contribute to the emission of greenhouse gases, and therefore to global warming, one of the greatest threats to freshwater resources' (Barlow, 2003, p22). Diversions where water is carried in giant pipes, Barlow informs us, will result in water depletion in the long term. Many poor rural areas become deserts when water is 'sold, expropriated or just plain stolen' (Barlow, 2003, p23). The final technology being used is very expensive and very energy intensive – desalination. It creates a lethal by-product and other contaminants which are pumped back into the sea. Furthermore, sewage is often dumped into the sea, which is then drawn back into the desalination plant for human consumption. The very new technologies such as Atmospheric Water Generators and Cloud Seeding are privately owned, allowing the companies to develop the market without competition.

Clearly, with the inevitable water crisis looming, water becomes a commodity worthy of value. However, the shift from a public to a private model is traced back, not too far, to Margaret Thatcher's Britain, and Ronald Reagan in the US. British publicly owned water was sold off to private companies at bargain prices. Millions had their water cut off. With the increasing neoliberal policy adopted by the World Bank of forcing developing countries closer to a market driven economy, water services in the Global South became targets for privatization.

Barlow explains how the World Bank and big water companies set out to promote a major shift in water policy, seeking buy-in of NGOs, think tanks, media and the private sector. 'Through its Water Capacity Building Program, the World Bank Institute (the capacity building arm of the bank that promotes bank values and programs through education and outreach) has put thousands through intensive programs on private water management' (Barlow, 2003, p42).

'For with the acceptance of water as a commodity comes the dilemma of what to do with the idea of water as a basic human right. In other words, if we are willing to use monetary value as our sole guiding principle for water extraction, treatment and distribution, on what grounds do we make moral decisions about how much water is enough and who is consuming too much? Just because someone can afford to pay the cost of filling their swimming pool or washing their cars every day should they have the right to do so when others are struggling to survive with no water at all? (McDonald in Francis, 2005, p 33)

Francis (2005) tells us that in 2001, South Africa adopted a policy of Free Basic Water (FBW), which aims to provide each household with 6000 liters of clean water every month free of charge.

This is deemed the minimum amount needed for survival. One of the strongest critics of FBW is that the quantity defined by the government is insufficient - 33 l per day compared with the UN suggested 50 l for a household. Also because it is per household – poorer families tend to have larger numbers in a household. Since 2002, there is a justiciable right to water. But what does this mean? It places an obligation on the government to take reasonable action to achieve these rights. The problem is that there needs to be a commitment of financial resources and cooperation amongst federal and local officials to achieve anything useful with this, and to date, this appears not to be the case.

Challenge Box: Do you think water should be free for all? Justify your response.

3.3   THE WATER WARS

while trade liberalization was perhaps the archetype disagreement on development strategy in the 1980s and 1990s in the 1990s and 2000s this role has been taken over by water privatization and the passions it arouses (Kanbur, 2007 p1)

The term 'water wars' referred originally to the famous protests in Cochabamba, Bolivia, in the year 2000, when the government sold the municipal water supply of the city to an international water company. There were many street battles and loss of life with huge debates worldwide. But what are the underlying reasons for disagreements on water, amongst those who proclaim it to be for poverty reduction? Kanbur's understanding (Kanbur, 2007) of why people might be against water privatization is that the selling of some things such as slaves and child labour is intolerable to them, and this must also be the case with water. However, there is a much deeper political divide in most of the papers you can find on the subject. Either you believe that the government should look after our basic services, or you believe that private industry should do this. In some cases, we find alternatives to this but not many.

Kanpur picks out a debate between Vandana Shiva and Fredrik Segerfeldt, which he quotes from extensively, and we will do so again as it illustrates the differences very well.

The World bank started to push the Delhi government to privatize Delhi's water supply...The contract between Delhi Jal Board... and the French company Ondeo Degrement (subsidiary of Suex Lyonnaise des Eaux Water Division – the water giant of the world), is supposed to provide safe drinking water for the city...On December 1, 2004, water tariffs were increased in Delhi. While the government stated that this was necessary for recovering costs of operation and maintenance, the tariff increase is more than ten times what is needed to run Delhi's water supply. The increase is to lay ground for the privatization of Delhis water and ensure profits for the private operators...the tariff increase is not a democratic decision nor a need based decision. It has been imposed by the World Bank. ....How will they give water to the thirsty? Cremation grounds, temples, homes for the disabled, orphanages which paid Rs 30 will now pay thousands of rupees. ...The World Bank driven policies explicitly state that there needs to be shift from the social perception to a commercial operation. This worldview conflict lies at the root of conflicts between water privatization and water democracy. Will water be treated as a commodity or will it be viewed as the very basis of life? The common argument for privatization and price increase is that higher costs will reduce water use. However, given extreme income inequities a tariff increase that can destroy a slum dweller or poor farmer is an insignificant expenditure for the rich. Privatisation as dictated by the Asian Development Bank / World Bank thus means that water will be diverted from the poor to the rich, from rural to industrial areas... Water privatization aggravates the water crisis because it rewards the waste of the affluent, not the conservation of resource of prudent communities. Sustainable and equitable use needs water democracy not water privatization.' Shiva, quoted in Kanbur, 2007

Figure 3.4. Vandana Shiva.

Ninety-seven percent of all water distribution in poor countries is managed by the public sector, which is largely responsible for more than a billion people being without water. Some governments of impoverished nations have turned to business for help, usually with good results. In poor countries with private investments in the water sector, more people have access to water than in those without such investments. Moreover there are many examples of local businesses improving water distribution. Superior competence, better incentives and better access to capital for investment have allowed private distributors to enhance both the quality of the water and the scope of its distribution. Millions of people who lacked water mains within reach are now getting clean and safe water delivered within a conventional distance…The main

argument of the anti-privatisation movement is that privatization increases prices, making water unaffordable for millions of poor people. In some cases it does, in others not. But the price of water for those already connected to a mains network should not be the immediate concern. Instead we should focus on those who lack access to mains water, usually the poorest in poor countries…They usually purchase water from small time vendors, paying an average of 12 times more than for water from regular mains and often more than that. When the price of water for those already connected goes up, the distributor gets both the resources to enlarge the network and the incentives to reach as many customers as possible. When prices are too low to cover the costs of paying new pipes, each new customer entails a loss rather than a profit, which makes the distributor unwilling to extend the network. Therefore, even a doubling of the price of mains water could actually give poor people access to cheaper water than before. True many privatizations have been troublesome. Proper supervision has been missing. Regulatory bodies charged with enforcing contracts have been non-existent, incompetent or too weak. Contracts have been badly designed and bidding processes sloppy. But these mistakes do not make strong arguments against privatization as such but against bad privatizations. Let us therefore have a discussion on how to make them better instead of rejecting them altogether. Greater scope for businesses and the market has already saved many lives in Chile and Argentina, in Cambodia and the Phillipines, in Guniea and Gabon. There are millions more to be saved. (Segerfeldt, quoted in Kanbur, 2007)

Figure 3.5. Fredrik Segerfeldt.

Barlow does not need a debate. She is convinced. 'Almost twenty years of documented cases of the failure of privatization and growing opposition to the World Bank and the water service companies in every corner of the globe have revealed a legacy of corruption, sky- high water rates, cutoffs of water to millions, reduced water quality, nepotism, pollution, worker lay-offs and broken promises' (Barlow, 2003, p58). But the main issue that Barlow has is that water companies are for profit 'the ultimate goal of private companies is to make profit, not to fulfill socially responsible objectives such as universal access to water. In countries where most of the population earns less than two dollars a day, notes Sara Grusky of Water Watch, private companies cannot meet shareholder obligations to provide a market rate of return. Nor can they expand their services to a population that cannot pay. The only way that the private sector can stay competitive in such a situation is to have access to public subsidies, the very thing they were brought in to relieve' (Barlow, 2003, p58).

3.4   WATER IN BUENOS AIRES

Figure 3.6. Shantytown in Buenos Aires. Many residents of the shantytowns do not have running water.

Buenos Aires is an often quoted example of the water wars debates. It is one of the largest concessions to date - the major shareholder being the French company Suez Lyonnaise des Eaux mentioned by Shiva above. Privatisations can be management contracts, lease contracts or full concessions to the private company. In our discussion below, the first case comes from a group of authors who are all associated with the World Bank and present the case for privatisation in Buenos Aires. The second group represents a group of academics in the north and south who work on what they call the 'Municipal Services Project.'

Abdala (1997) suggests that prior to reform, water and sewerage services provided by Obras Sanitarias de la Nacion in Buenos Aires suffered from a lack of investment and inadequate maintenance.. 'overstaffing, unresponsive customer service, high levels of unaccounted for water (UFW) and low collection rates. Privatisation on the other hand brought about positive changes, resulting in rapid improvements in the performance of the company now called Aguas Argentinas' (Abdala, 1997, p4).

He concludes that the government came out as a loser but that employees in turn became beneficiaries as they received ownership of 10% of AA shares. Abdala states 'the result applies to those who remained employed and dividends could not be expected sooner than 1999. For this, workers who entered into the voluntary retirement programmes we assume that the severance payments that they received were enough to compensate for their disutility from being laid off. Buyers also 'came out with welfare gains' and consumers 'in turn are the big winners.' Existing customers received the largest share of this pie as there were important cuts in the average price for usage. New users also benefited as privatization reduced rationing by increasing access. Consumers also received benefit through improved quality effects and public health that we were unable to quantify. On the quality side, this is clearly the case of improved water pressure levels waiting time for repairs and customer service' (Abdala, 1997, p25).

Furthermore, Galiani et al. (2005) found that child mortality fell 5-7 % in areas that privatized their water services overall and that the effect was largest in the poorest areas. 'In fact, we estimate that the privatization of water services prevented approximately 375 deaths of young children per year' (Galiani et al., 2005, p1). They state in their conclusions that their results shed light on two important debates. The first they describe as follows 'One concern that has postponed privatization of water systems around the world is the fear that private operators would fail to take into account the significant health externalities that are present in this industry and, therefore, under- invest and supply sub-optimal service quality. Contrary to this concern, we find that the effect of privatization on health outcomes has been positive. Private operators have accomplished the network expansion and quality requirements specified in the privatization contracts or at least their level of attainment has been superior to the performance under public management. While the private sector may be providing suboptimal services they are doing a much better job than either the public sector or the non profit cooperative sector' (Galiani et al., 2005, p28).

The second area is described equally strongly. 'There is a growing public perception that privatization hurts the poor. This perception is driven by the belief that privatized countries raise prices, enforce service payment and invest only in lucrative high income areas. Instead, we find the poorest populations experienced the largest gains from privatisation in terms of reduction in child mortality. Privatisation appears to have had a progressive effect reducing health inequality' (Galiani et al., 2005, p28).

Some groups acknowledge that there were problems but do not believe that privatisation itself is to blame. The signing of a concession contract for the Buenos Aires water and sanitation system in December 1992, attracted worldwide attention, and caused considerable controversy in Argentina. The concession was implemented rapidly and, according to Alcazar et al. (2000), reform generated major improvements in the sector, including wider coverage, better service, more efficient company operations, and reduced waste. 'Moreover, the winning bid brought an immediate 26.9 percent reduction in water system tariffs. Consumers benefited from the system's expansion and from the immediate drop in real prices, which was only partly reversed by subsequent changes in tariffs, and access charges. And these improvements would probably not have occurred under public administration of the system.' Still, the authors show that 'information asymmetries, perverse incentives, and weak regulatory institution' could threaten the concession's sustainability. 'Opportunities for the company to act opportunistically - and the regulator, arbitrarily - exist, because of politicized regulation, a poor information base, serious flaws in the concession contract, a lumpy and ad hoc tariff system, and a general lack of transparency in the regulatory process. Because of these circumstances, public confidence in the process has eroded. The Buenos Aires concession shows how important transparent, rule-based decision-making is to maintain public trust in regulated infrastructure' (Alcazar et al., 2000).

McDonald and his group, on the other hand, take the opposite perspective (Loftus and Mcdonald, 2001). They explain that President Menem rushed through the national Administrative Reform Law declaring a state of economic emergency with regard to the provision of public services. The law authorized the 'partial or total privatization or liquidation of companies, corporations, establishments or productive properties totally or partially owned by the state, including as a prior requirement that they should have been declared subject to privatization by the Executive Branch, approval for which should in all cases be provided by a Congressional law. Through such a decree, Menem was able to privatize the Buenos Aires water and sewerage network Obras Sanitaras de la Nacion (OSN) without pubic consultation, arguing that it was 'urgent' to press on with reforms' (Loftus and Mcdonald, 2001, p11). Even amongst union leaders, there was a feeling of inevitability and that fighting was futile. The hyper-inflation which happened during this time in Argentina 'essentially disciplined the population into accepting privatization as a solution' (Loftus and Mcdonald, 2001, p12).

Table 3.1 shows the coverage at the start of the concession in 1993 [reproduced from (Loftus and Mcdonald, 2001, p14)].

Of the 30% population not connected to the water network in 1993, 95% obtained water from wells with pumps, and for those with sewerage services, 88% disposed of this through cesspools and the rest directly into rivers.

Table 3.2 shows the five yearly performance targets [reproduced from (Loftus and Mcdonald, 2001, p18)].

By 1999, Aguas Argentinas claims that water coverage had reached 82.4% and this meets the performance targets. However, they were at 61% for sewerage. Sewerage infrastructure had not kept pace with water delivery expansion and it is claimed that this is because water delivery is twice as profitable as sewerage for Aguas Argentinas. 95% is still dumped into the Rio Del Plata. We have

Table 3.1: Water coverage before the concession in Buenos Aires (Loftus and Mcdonald, 2001)	
Water System	
Number of Connections	1,170,000
Average Production (m^3/day)	108,950,000
Treatment Capacity (m^3/day)	3,640,000
Length of Water Pipe System (kms)	11,000
Total Population	8,580,000
Served Population	6,000,000
Sewerage System	
Number of Connections	700,000
Served Population	4,700,000
Volume Collected (m^3/month)	82,232,000
Volume Treated (m^3/month)	3,413,000

Table 3.2: Five yearly performance targets (Loftus and Mcdonald, 2001)

Year of conces.	% population served by water	% population served by sewerage	% Collected sewage treated		Network renovation cumulative %		% unacc. for water
			Primary	Secondary treatment	water	sewage	
0	70	58	4	4	0	0	45
5	81	64	64	7	9	2	37
10	90	73	73	14	12	3	34
20	97	82	82	88	28	4	28
30	100	90	90	93	45	5	25

been told by the locals that the problems with privatisation follow from the fact that the higher water tables are now completely polluted as a result of the untreated sewage.

Challenge Box: Try to find out when and if water in your city/country has been privatized. What are some of the popular believes about the costs and benefits of this?

Figure 3.7. Rubbish tips and water ways in Buenos Aires. (The bicycles line up as cartoneros (see the chapter 'Throwing Away My Rubbish') wait until they are allowed into the landfill site to collect anything they might be able to sell.)

Apart from service, the main arguments for privatization are cost and accountability. It is usually said that private companies are more efficient than the public sector and reduce costs to the end user. Here, Loftus and Macdonald believe claims have been exaggerated. Costs went down 27% and then went up 20% - critics apparently argue that the higher prices were made just before so as to make the company look good. 'The effect of the increases (prior to privatization) was to allow the company to offer what seemed to the public to be a 27% decrease in costs, even though in reality it was a manufactured reduction' (Loftus and Mcdonald, 2001, p19).

It is also said that privatisation generates better public accountability. In fact, Loftus and Macdonald suggest that the opposite has happened. 'Starting with a Presidential decree in 1989 which unilaterally declared that the city's water and sanitation would be run by the private sector, all decisions about the extent and scope of privatization in these sectors were made behind closed doors. There was no public debate on the matter, and the first (and only) public consultation did not take place until June 2000, seven years after the concession had begun' (Loftus and Mcdonald, 2001, p3).

Loftus and McDonald summarise by stating that 'although there have been some impressive gains in the extension of water infrastructure, the majority of the concessions' negative impacts have been mostly felt by the poorest sections of BA. Many poor households have fallen into serious arrears and have been disconnected from the network, especially prior to 1998. ...environmentally, those living in the poorest areas of BA have also been faced with the negative effects of rising groundwater and the health risks associated with nitrate contaminated aquifers. These municipalities have some of the lowest average incomes in the Greater BA and yet a large part of the financial burden for extending the network has fallen on these households' (Loftus and Mcdonald, 2001, p29).

As Kanbur (2007) puts it, 'Two more opposing descriptions of outcomes could hardly be possible. According to one analysis, the privatization was an unmitigated disaster. According to another, it saved children's lives' (Kanbur, 2007, pg8). These views, he says, are taken by two general tendencies which he calls the Civil Society tendency (CS) and the Finance Ministry tendency (FM). Kanbur gives away his preference by suggesting that the FM framework is more generally open and more humble about their findings and frameworks (Kanbur, 2007, pg11); however, his key interest is to 'understand the different methods of analysis and different ways of formulating questions. After a few such meetings, we would be in a position to judge whether the divide can be bridged' (Kanbur, 2007, pg11).

On March 21st 2006, the Argentina Government rescinded the thirty year contract of Argus Argentinas. An April 2007 report by the city's ombudsman stated that most of the population of 150,000 in the southern district of the city lived with open-air sewers and contaminated drinking water (Barlow, 2007, p106). Suez was also forced to abandon its last stronghold in Argentina, the city of Cordoba, when water rates were raised 500% on one bill' (Barlow, 2007, p107).

3.5   WATER IN AFRICA

We see a similar situation in studies of water privatization in Africa although it has been slower to develop. By the end of 2000, 93 countries had privatized some of their water services but only 7 projects in the Middle East and North Africa and 14 in sub-Saharan Africa. Kirkpatrick et al. (2004) suggest that across Africa there is better performance in private utilities compared to state owned utilities, but there are found to be no statistically significant cost differences. They analyse reasons why water privatization may prove problematic in lower-income economies but draw no conclusions - stating that the results are not significant. More poignant is the final statement, however: 'Finally it needs to be stressed that while the paper has concentrated upon a number of performance measures, a more comprehensive study would take account of possible effects beyond those discussed. For example, we have seen that privatization tends to lead to more water metering, but what is the impact of this on water consumption and health? Around major cities in developing countries lie shanty

Figure 3.8. Usual way of collecting water in Mamathe, Lesotho.

towns populated with squatters and others without legal property rights. How are their interests served by water privatization? Water privatization usually means the involvement of a handful of major international companies, but what effect does this have in terms of developing indigenous ownership of socially important assets? Also, if privatization leads to full cost recovery in water, is this outcome compatible with poverty reduction and what are the environmental implications of privatization? Clearly, water privatization raises a complex set of considerations that deserve fuller exploration than has been possible here because of data limitations' (Kirkpatrick et al., 2004, p24). It seems that some of the water war issues and huge differences opinion are exacerbated by the fact that researchers are measuring different things and, in some cases, not the ones which are relevant to issues of social justice.

One paper, which does address this directly is entitled 'Water Justice in Africa: Natural Resources Policy at the Intersection of Human Rights, Economics and Political Power' (Francis, 2005). In this paper Francis analyses water as a social justice issue in South Africa. She explores historical changes in South African water law. Francis points out that the remarkable part of South Africa is its inequality despite being post apartheid because of years of colonial and fifty years of apartheid rule. White South Africans, who comprise just 10% of the population, predominantly live in conditions similar to those in wealthy nations. However, she states that one third of the population live in conditions of less than U.S. 2$ per day with 30-60% unemployment amongst the blacks. All white suburbs have 50% of the residential water use and only 27% of black households have running water compared with 96% white. Insufficient and inadequate infrastructure for delivering water services to poor communities is compounded by scarcity in water. Much of this, Francis suggests, goes back to apartheid. Black citizens were prohibited from residing in urban areas and could only have legal residence in crowded 'homelands.' They were forced to live in areas with poor soil and limited water, to clear out unwanted shantytowns and make room for commercial forestry etc. on valuable land. First and third world economies developed independently of one another ... 'Current discrepancies in the allocation of water can be traced back to apartheid which reserved the resource for the landed white minority through convoluted ... laws linking water rights to land ownership.' Most water was used inefficiently and virtually for free by large scale commercial farmers – a 'dominant group' with 'privileged access to land, water, and economic power' (Francis, 2005, p8).

In 1990, sitting president Frederik Willem de Klerk announced the end of apartheid and released from prison Nelson Mandela, the leader of the African National Congress (ANC). In 1994, ANC came into power. Today, there is a Constitution that embraces human rights' principles and contains a Bill of Rights, which enshrines rights to many basic services including water. However, says Francis, within two years, they had transformed into neoliberal policies aimed at macroeconomic growth and restrictions on public spending – lauded by the IMF. In such a system, she explains, it is the poorest who disproportionately pay the price for economic restructuring. 'The country has not realised significant foreign investment but instead has shed hundreds of 1000s of jobs primarily in sectors like agriculture and textiles which tend to employ poor black women' (Francis, 2005, p13) 5% of the population controls 80% of the countries wealth.

The Water Services act of 1997 codified the Constitutional right to basic water and sanitation bringing piped water within 20 metres of each household. The national water act (NWA) of 1998 is widely considered to be one of the world most progressive water policies on paper, and it disconnects water rights from land ownership. The NWA was widely supported by black Africans but opposed by business interests, white farmers and outgoing National party – however, according to Francis, it creates loopholes that allow the status quo to prevail and facilitates the implementation of neoliberal financial policies.

'Insufficient access to potable water was a widespread problem before the ANC came into power, but to the extent that water was available, it was provided for free or at a highly subsidised price.' The government then introduced cost recovery and the price of water increased dramatically and the connection fees and volumetric charges proved too costly for low income households who had been getting their water from communal taps. 'As a result many individuals were forced to resort to collecting water from streams, canals or stagnant puddles' (Francis, 2005, p25).

She quotes Metolina Mthembu, a 70 year old resident in Mbabe village in KwaZulu-Natal province, 'who is glad the new government installed a tap outside her home two years ago. But now the water costs money and people here are poor. There are no jobs. We must choose between food and water, so we buy food and pray that water does not make us ill. It is a bad gamble. Many, many of us have grown sick from the water' (Francis, 2005, p25).

There is now a substantial debt among low income families due to water service charges, which according to one national survey averaged U.S. 290 by 2001. Amazingly, there seems to be a debate about why there is a non payment of water services and suggestions about a culture of non payment. Water cut offs are the most common response to inability to pay. More than 10 million have had their water disconnected. Meantime, local governments, Francis says, are trying to turn water services into profitable ventures in order to attract private investors, encouraged by the IMF and World Bank.

'Lyonnais des Eaux has come knocking on my door on two occasions. These French water companies have become too powerful to resist. The takeover is inevitable. I want to run our services like solid business units to make sure we negotiate from a position of strength when it does happen'. (Francis,2005, p12).

Challenge Box: If your city has recently experienced privatization of water, try to find out who the major shareholder is in the company that runs the organisation. Locate any news articles that you can about this company and consider the various viewpoints expressed.

3.6   PRIVATISATION

We have been discussing privatization of water and we might assume what this means. It is worth, however looking more deeply at the various definitions of this phenomenon.

Artists representation of a stunt organised by the UK National WDM (World Development Movement) and carried out by Brighton and Hove, UK World Development Movement at the 2006 Labour Party Conference. It was to highlight WDM's Dirty Aid Dirty Water campaign to protest against the UK Development organisation's use of aid money to pay private consultants to push water privatisation on developing countries rather than supporting improvement of public utility provision.

Figure 3.9. Water for sale on Brighton Beach.

'in its narrowest sense , privatization happens when the state sells its assets to a private company, along with all of the maintenance, planning and operational responsibilities that these assets entail. Over the past 30 years states have divested themselves of airlines, railroads, telephone services, health facilities and other services, thereby unlocking a new phase of capitalist expansion and innovation. Divestiture, as this form of privatization was the model of privatization adopted in the UK under Margaret Thatcher in the late 1980s with entire systems of public services delivery being sold to private firms'(McDonald and Rulters, 2006 p9)

The South Africa Workers Movement (SAMWU), says that privatization takes place when (Samson, 2003):

- The government contracts private companies to run certain parts of a service. This they call outsourcing.

- The government gets a private company to manage a government department or unit. This is usually called a management contract.

- The government gets community groups to do work that used to be done by the municipality.

- Government departments are changed into private companies which are owned by government. In Jo'burg these private companies are called utilities, agencies and corporatised entities or UACs.

- Government departments are changed into business units, which are completely separated from other departments in the municipality. These units are still owned by government but they operate like private businesses with the same kind of profit incentives (Samson, 2003, p13).

McDonald and Rulters (2005/6) explain the different forms of public private partnership in full in Table 3.3.

Table 3.3: Different forms of public private partnership (McDonald and Rulters, 2005/6)	
Full Divestiture	Divestiture refers to a situation where a public utility or service has been fully privatized.
Service Contract	This is the least risky of all partnership types. The public authority retains responsibility for operation and maintenance of the service - components are contracted out. 1-2 years
Management Contract	The management contractor operates and maintains services or parts of services. 2 - 20 years
Lease or Affermage	The lessor rents the facility from the public authority which transfers complete managerial responsibility for operating and maintaining the system to a private company
Concession	Investment linked contract. Concessionaire has responsibility for services and capital investments. Ownership of fixed asses is assigned to the local authority at the end of the contract. 25-30 years
BOOT	Build Own Operate and Transfer contracts - new parts of a service system. 25 years
Community/NGO Provision	Transfer of some or all of the responsibility for service provision to end user or not-for-profit intermediary. Common in low income urban settlements. Women often carry the burden of this labour.

Sometimes we confuse terms such as privatization, corporatisation and commercialisation. Macdonald helpfully teases out the differences. Commercialisation, he suggests, refers to a process by which market mechanisms and market practices are introduced into the operational decision making of a public service. A popular institutional form of commercialisation is corporatisation where services are 'ringfenced' into stand alone business units owned and operated by the state but run on market principles. The link between corporatisation and privatization is that there has been a change in the management ethos to an increasing focus on a narrow short term financial bottom line. This means that public enterprises can be even more commercial that privatized counterparts. Corporitisation also promotes outsourcing as an operating strategy and cost cutting, and it can act as a gateway for direct private sector investment.

'Tying all this together are the underlying processes of commodification. Only when public services are treated as a commodity can they be effectively commercialised and eventually privatised. It is at this politico-economic juncture that we see the full significance of such as transformation emerge e.g., water stripped of its image as an abundant nature provided good for public benefit, to water as a scarce monetised entity subject to the same laws and principles of market as shoes, lampshade or computers' (McDonald and Rulters, 2005/6, p13).

3.7   WHY IS PRIVATIZATION TAKING PLACE?

Neoliberal analysts have argued that privatization occurs because states fail: state officials are rent seeking, inefficient, unaccountable, inflexible and unimaginative. Privatisation is seen as a rational and pro poor policy choice, obvious to anyone willing to look at track records of public versus private debate. McDonald and Rulters quote Hodge (McDonald and Rulters, 2005/6, p15) who identifies five core theories: public choice theory, agency theory, translation costs analysis, new public management and property rights theory – they all use the assumption that people respond best to market incentives and that market based systems are inherently more efficient. Macdonald et al. argue by contrast 'that the privatization of public services has not happened because it has been inspired by some renewed sense of cultural enthusiasm for the market but rather that it has become a necessity imposed on the state by economic circumstances' (McDonald and Rulters, 2005/6, p16). They state that it is possible to trace the shift with water. Economies and construction of new dams shrank in 1970s and competition for opportunities intensified. The state could not support spending on new infrastructure. For the water and engineering construction industry, 'privatization was seen as a way to absorb idle productive capacity and excess commodities' (McDonald and Rulters, 2005/6, p16).

They say that nowhere in the mainstream literature is there a recognition of the argument that privatization and commercialization are a response to the pressures of an ever expanding marketisation of social relations under capitalism. Nor is there any discussion in mainstream debates of the radical thesis that capitalists must constantly seek new geographies and sectoral areas of investment as a response to a capital overaccumulation or that capitalists are constantly forced to recreate the physical means of production of built environments that facilitate market expansion. For global capital seeking new areas, they go on to say ' the public sector provides an enticing opportunity. Water is particularly attractive' (McDonald and Rulters, 2005/6, p16).

Barlow, however, maintains that strong civil resistance is the key to the retreats that have been seen so far. 'Today a coordinated and highly effective international water justice movement is fighting both the power of the private water companies and the abandonment by their governments of the responsibility to care for their national water resources and provide clean water to their people' (Barlow, 2008, p124).

3.8   FINAL THOUGHTS

In this chapter, we have presented strong opinions for and against the issue of the privatization and commodification of water. You may already be paying for your water bills and know how expensive they can be. Hopefully, we have given you enough food for thought to allow you to consider the movements of your own government on this issue, but also to help you think through some of these perspectives if you begin to work for an engineering company who is responsible for carrying out the decisions made about this at a political level.

Challenge Box: Consider the pros and cons of privatization of public services after all the various discussions in this chapter.

CHAPTER 4

Awakened by an Alarm Clock

Figure 4.1. Awakened by an alarm clock.

In this chapter, we are focusing on time. As a citizen, we wait for time to pass or we want to try to beat time. As an engineer, many of our processes are controlled by time. Looking at time from the perspective of social justice causes us to ask different questions.

4.1   ENGINEERING AND OUR CONCEPT OF TIME ARE IN-TRICATELY RELATED

The alarm clock over there on the nightstand, nestled among the books, magazines and cough drops, that lies silently, stealthily in wait to roust you from the deepest of sleeps, is one of the most important symbols or metaphors of our modern, technologically advanced society. The clock, in many ways rules our lives, telling us when to wake up, when to get off to class, when to meet a friend, really when to do just about anything! In this next section, we will explore how modern society came to be governed by this merciless ruler and speculate on how things might be different if we adopt a new understanding of time. This will be true for all professions, but the resultant consequences for engineering may be the greatest of all.

By the end of this section it is hoped that you will be able to:

- Describe the importance that *time* plays in modern, technological society.

- Describe the links that exist between our concept of time and our most commonly held scientific paradigm, i.e., *nature as a machine.*

- Describe the relationships that exist between *nature as a machine* paradigm and other academic disciplines.

- Understand a new scientific paradigm, which has superseded our notion of nature and the universe as mechanical.

- Appreciate how the new paradigm may change our understanding of time.

- Examine the possible future implications of this new understanding for engineering.

- Elucidate possible consequences for this new understanding in other disciplines.

> Time present and time past
> Are both perhaps present in time future,
> And time future contained in time past.
> If all time is eternally present.
> All time is unredeemable.
> T.S. Elliot.

In *Technology and Society*, we pointed to the connections that exist between technology, engineering practice and the society in which it operates. As was discussed, engineering practice and the development of technology operate as systems of thought and practice, which provide a framework for what, we imagine and what we wish to create. Different sets of values in a society could lead to an entirely different set of technological developments.

In the present chapter, we will focus on the alarm clock, that ubiquitous and ever-present feature of modern life. More than simply a collection of gears and levers or circuit boards and LED's, the alarm clock is symbolic of many important aspects of modern life, has significance as an indicator of our societal values, and foreshadows a future for both our society and our planet. We will look back at different technological developments and societal changes that led up to the alarm clock and our conceptual model of time. We will then consider the consequences that our present model for time has in both the short term and long term. Lastly, we will imagine a very different model of time and explore the possibilities that might result.

> **Challenge Box:** Think of five ways in which the clock governs your behavior during the day. For each way you have identified, is there some connection with an important aspect of modern society?

4.2   THE ALARM CLOCK

A **clock** (from the Latin *cloca*, "bell") is an instrument for measuring time and in its modern form dates back to approximately the 15th century. Adding the alarm to the clock changes the device from one that simply measures the quantity *time* to one that exerts control over our lives or at least attempts to do so! Today, we move from doctor's appointment to dentist's appointment, from one classroom to the next with precision that rivals that found in a highly trained military organization. When we need to travel via airplane, train or bus we consult a schedule which provides us arrival times and departure times, sometimes even accurately. We arrange our evenings and weekends around the posted times of various entertainment outlets from movies and plays at our nearby theaters to situation comedy TV shows to kickoff times of Super Bowls and World Cups. On some occasions, you may even wish to consult with that favorite professor during her/his announced office hours with a reliance again on the universality of *time*. Everything, it seems is governed by what we have come to regard as the phenomenon known as *time*.

As engineers, the present day understanding of time is integral to the practice of our profession. Engineers model a wide range of phenomena, postulating equations, which predict behavior. We describe the resulting solutions as either steady state or unsteady or transient, depending upon whether those solutions are independent of time, change with time without ever reaching a steady state value, or change with time until eventually a steady state value is reached. In fact, engineers today think of *time* as a fourth independent variable, comparable to the three independent variables we typically associate with three-dimensional space. As engineers, we are often tasked with solving problems that have never been solved before using our design methodologies, integrating analytical and creative thinking skills. In design, we develop milestone schedules to layout and track the various tasks that will be required to reach the desired end goal. Whatever design strategy we chose, it is time that serves as the underlying concern. Our final design and product must be delivered at a certain specified and agreed upon time.

It is the clock that tells us the appropriate time. Yet, the clock, particularly the mechanical clock with or without the bells and whistles necessary to sound the alarm, has only been in widespread use for the last several centuries. We shall first consider the importance of time in present day society. Then in this next section, we will look back in time to consider one important technological advancement, the mechanical clock, that leads up to our conception of time today. We shall then imagine the role of time in the future and consider the implications of our present understanding for that possible future.

4.3 MODERN SOCIETY AND WELCOME TO THE MACHINE

Welcome my son, welcome to the machine.
Where have you been? It's alright we know where you've been.
You've been in the pipeline, filling in time, provided with toys and Scouting for Boys.
You bought a guitar to punish your ma,
And you didn't like school, and you know you're nobody's fool,
So welcome to the machine. (Pink Floyd, *Welcome to the Machine*.)

Modern engineering has been intricately linked to the design, development and manufacture of machines. As a society, we expect our machines to run as "clocks," that is, we wish our designed products to operate dependably, with repeatability and preciseness. The clock is often used as a metaphor for modern industrialized society, and until very recently, we were often times described as living in the "machine age." Our modern, highly mechanized civilization is the outgrowth of the design and development of countless engineering devices as well as a combination of numerous habits, ideas customs and cultural norms. Even the universe itself is considered by many in technological fields to be the ultimate machine whose behavior is understandable and thus predictable *when*, not *if*, science finally discovers the proper set of equations or as physicists refer to it, the *theory of everything* (Hawking, 1998).

As you walk about the common grounds of your college campuses, you are likely to hear bell chimes emanating from the campus campanile or clock tower. Those tones tell you the time of day and whether or not to walk leisurely or sprint at top speed to your next class. The universally recognizable pattern of tones breaks up the day into twenty-four hours and typically each hour into fourths or 15-minute intervals. The bell towers date back to the 13th and 14th centuries in Western Europe, the most famous being the clock built by Heinrich von Wyck in Paris. With the advent of the clock and its ever imposing and obvious presence, time soon took on a very different character then it had earlier. According to Mumford, "Time took on the character of an enclosed space: it could be divided, it could be filled up, it could even be expanded by the invention of labor-saving instruments" (Mumford, 1963). Engineering soon became tasked with designing, developing and delivering instruments or devices, which would enable the user to save time and thereby do more.

Engineers also became concerned with designing and implementing process that would save time as well leading to the field of time and motion studies or time-motion study. A time and motion study is used to develop a new process or series of sequences/steps to reduce the number of motions in performing a task in and increase productivity. The best known experiment involved bricklaying. Through carefully scrutinizing a bricklayer's job, Gilberth in the 19th century reduced the number of motions in laying a brick from 18 to about 5. The focus was on laying more bricks in the same amount of time and thereby expanding the time, bringing mechanical efficiency through coordination and careful design.

In many cultures, particularly in the West, efficiency was soon seen as an important and highly desirable trait. This attribute, the ability and the desire to do more and more in the same period of time, is profoundly evident today. While only a few years ago, it was sufficient to drive a car from one place to the next, now we find it necessary to talk on cell phones, text messages, use instant messenger, and countless other tasks all from behind the steering wheel. Restaurants are no longer settings for meals and casual or intimate conversations but rather a place to eat quickly, and catch all that is happening in the world while glancing at monitors tuned to stations focused on entertainment, the weather, news and sports. Our homes, dormitories, offices are all filled with all manner of gadgets and devices (all designed, developed and delivered by engineers) that are sold to us as items that will allow us pack more and more into the same 24 hour period. Perhaps, the motto that best describes the modern world is this: Just give me more time! I need more time! (Tolle, 2004).

4.4   THE ASSEMBLY LINE

An examination of modern *assembly lines* provides insight into how important the concept of time has become. An *assembly line* is a manufacturing process in which interchangeable parts are added to a product in a sequential manner to create an end product. Until the 19th century, a single craftsman or team of craftsmen would create each part of a product individually, and assemble them together into a single item, making changes in the parts so that they would fit together. This linear assembly process, or assembly line, allowed relatively unskilled laborers to add simple parts to a product. While originally not of the quality found in hand-made units, designs using an assembly line process required much less training of the assemblers, and, therefore, could be created for a lower cost (Mumford, 1971). Modern assembly lines often have much more complicated interdependencies with not only workers and but also robots and other devices linked in the desire to produce more in less time at a lower cost. Henry Ford was one of the first to apply assembly line manufacturing to the mass production of affordableautomobiles. This achievement not only revolutionized industrial production in the United States and the rest of the world, but also had such tremendous influence over modern culture that many social theorists identify this phase of economic and social history as "Fordism" (Watts, 2005).

Figure 4.2. Assembly line in Stolen water company.

An assembly line is a manufacturing process in which interchangeable parts are added to a product in a sequential manner to create a finished product. Hounshell (1984)

While some scholars credit Henry Ford with the creation of the middle class in the United States, soon other consequences of the assembly line became apparent and still exist today. Many workers soon became unhappy with the assembly line. They felt alienated from the products of their work as they often never had the satisfaction of seeing the finished product. Workers in assembly lines can easily be thought of as simply another cog in the production machine and easily replaced. Because workers had to stand in the same place for hours and repeat the same motion hundreds of times per day, they often suffered from what are now called repetitive stress injuries. In addition, workers were often times exploited and forced to work in dehumanizing conditions.

Challenge Box: Think of any assembly lines you encountered today. What was your reaction? Imagine you were working in an assembly line of some kind. How do you think you would react?

4.5   NATURE AS MACHINE

Not only has the production of goods and the role of workers in that production been seen to be as a *machine* but so too has our conception of the natural environment. Our environment today is seen by many to be a collection of *natural resources* which we can use to satisfy our needs and desires. We also speak of maintaining the balance of nature or returning nature to its balance. We imagine we can do this by adjusting a whole range of input parameters and boundary conditions much the same way we design assembly lines to maximize product output. Nature is seen as governed by the laws of *causal determinism* (Bunge, 1963) in this view.

Causal determinism is based upon the belief that every effect has a cause and thus science, if clever enough and developed enough, can be used to explain all natural phenomena.
The stanford Dictionary of Philosophy,2009

Even if there is only one possible unified theory, it is just a set of rules and equations. What is it that breathes fire into the equations and makes a universe for them to describe? The usual approach of science of constructing a mathematical model cannot answer the question of why there should be a universe for the model to describe. Why does the universe go to all the bother of existing?

In causal determinism, the only thing that can be said to exist is matter and all things are composed of matter, and all phenomena are the result of matters' interactions. Our role as engineers then is to use this information discovered by scientists to help us to design devices and/or processes that meet our clients' needs. Ultimately, we envision nature or the universe to be a machine and we are the master mechanics whose goal is to first conquer and then modify the environment to suit our interests. Greene (1999).

There are grounds for cautious optimism that we may now be near the end of the search for the ultimate laws of nature.

Challenge Box: Do you share the view that the universe is material or matter only? What makes you feel confident in your view? How did you arrive at your view?

The consequences of the *Nature as machine* model (Botkin, 1993) have become more readily apparent than ever before. Plant and animal species are becoming extinct at a rate unsurpassed at any other time except for the mass distinction associated with the dinosaur disappearance. The oceans of the world are succumbing to years upon years of over fishing and pollution of various guises (Mastny, 2005). The Earth's climate is changing with the average land and ocean surface temperatures rising faster than ever before in recorded history. This phenomenon, referred to as *Global Warming*, confronts us with serious questions about the sustainability of our life styles and through extension ultimately our personal value system (World Watch Institute, 2006).

"An increasing body of observations gives a collective picture of a warming world and other changes in the climate system."
Intergovernamental Panel on Climate Change - IPCC (2001).

There are many other consequences that arise from the adoption of nature as a machine model. Some of these consequences are more explicit than others but all are equally directly resulting from the mechanistic conceptualization of nature. The suggestion that nature is a machine also suggests two additional points. First, some of us, specifically engineers, design machines to do our bidding. If nature is a machine then we can certainly manipulate (or at least attempt to do so!) the forces of nature to meet our needs and desires. We often speak of conquering and/or taming nature. While no mechanist would suggest that at the present moment in history, we can conquer/tame hurricanes, we do feel dangerously confident about having our wishes. We need look no farther than the devastation wrought by Hurricane Katrina, which struck the Gulf Coast of the U.S. in 2005, or the tsunami (Unesco, 2005) that devastated nations from Southeast Asia to the eastern coast of Africa with thousands of lives lost.

Challenge Box: What is your reaction to the conceptualization of nature and the universe as a machine?

4.6   ECONOMICS AND THE MACHINE

Once the "common sense" view that nature and the universe are deterministic and governed by a set of unchangeable laws, it was not long before other academic disciplines outside of the pure sciences were modeled in similar ways. This was particularly true in the field of economics and, not surprisingly, the *mechanical* view of economics still dominates today. The economist equivalent to the natural laws of physics put forward by Newton and his contemporaries is the *Law of Supply and Demand* (Smith, 2003). The law, originally developed by Marshall (1997), attempts to describe, explain and predict changes in the price and quantity of goods bought and sold in competitive markets. The law remains as a basic building block for a wide range of modern, more detailed economic theories and is an explanation of the mechanism by which many resource allocations are made. Note the use of the word *mechanism*!!

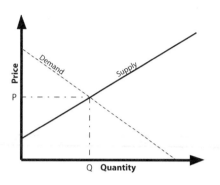

The Law of Supply and Demand is one of the most fundamental models of some modern economic schools, widely used as a basic building block in a wide range of more detailed economic models and theories.

If the natural laws of physics gave us the mechanical universe, what then did the laws of economics give us? Most likely, the answer would be capitalism. Commonly, the concept of capitalism refers to an economic system in which the means of production are primarily privately owned and operated for profit, with the investment of money or capital also determined privately, and decisions regarding the production, distribution and prices of goods, services and labor influenced the laws such as the law of supply and demand. As mechanists refer to the behavior of the natural world in harmony [i.e., reminiscent of *the music of the spheres* in physics (James, 1993)] with immutable natural laws, capitalists point to "the natural harmony of the rational self-interests of all men under capitalism – of businessmen and wage earners, of consumers and producers, of men of all races and of the competition of all levels."

4.7   LOOKING FORWARD IN TIME: A NEW PARADIGM FOR SCIENCE AND TECHNOLOGY

Science no longer conceptualizes nature and the universe as a machine with predictable patterns of behavior in the same manner as a mechanical clock. Today, we speak of self-organizing principles, and emergent properties (Bak, 1996). Self-organizing principles govern a process in which the internal organization of a system, normally an open system, improves automatically without being guided or managed by an outside source. Self-organizing systems typically display emergent properties. Emergent properties are properties that are dynamic in nature, that is, they change in time, and grow in complexity. Ultimately, the resultant structures dissipate or dissolve in time and the process begins anew. An emergent property can appear when a number of simple entities or agents operate in an environment, forming more complex behaviors as a collective. Examples of such properties include the construction of an ant hill or a bee hive wherein ants or bees come together and build complex structures that we as engineers often seek to emulate! Nature is thus not a collection of pieces and parts that can be replaced at our discretion. Once nature starts a process or we impact the starting of that process, we have no control over its final disposition. We no longer can speak of taming or conquering the natural world. That simply isn't in the cards!

We will have to see that we are the natural expressions of a deeper order. Ultimately, we will discover in our creation myth that we are expected after all.

Implicit in the new paradigm is a greater reliance upon community. It is not the individual ant or bee that matters but the entire population of ants and bees. A second implication of this new paradigm is that very simple actions ultimately may lead to very complicated and unforeseen results. Perhaps nature plays its own version of the *Rube Goldberg* game (Wolfe, 2000).

Figure 4.3. Rube Goldberg machine.

Rube Goldberg gets his think-tank working and evolves the simplified pencil-sharpener. Open window **(A)** and fly kite **(B)**. String **(C)** lifts small door **(D)** allowing moths **(E)** to escape and eat red flannel shirt **(F)**. As weight of shirt becomes less, shoe **(G)** steps on switch **(H)** which heats electric iron **(I)** and burns hole in pants **(J)**. Smoke **(K)** enters hole in tree **(L)**, smoking out opossum **(M)** which jumps into basket **(N)**, pulling rope **(O)** and lifting cage **(P)**, allowing woodpecker **(Q)** to chew wood from pencil **(R)**, exposing lead. Emergency knife **(S)** is always handy in case opossum or the woodpecker gets sick and can't work.

4.8 SELF GOVERNING PRINCIPLES, ECONOMICS, AND COMMUNITY

Self-organization is also relevant in chemistry, and is central to the description of biological systems, from the sub-cellular to the ecosystem level. Self-organizing behavior found in the literature of many other disciplines, both in the natural sciences and the socila sciences such as economics or anthropology.

Figure 4.4. Tennyson (2009)

In economics, according to some economists, the theoretical free market is self-organizing. Other economists have offered a completely different view. Bogdanov, scientist, philosopher, economist, physician, novelist, poet, and Marxist revolutionary, suggested that all human, biological and physical activities can be unified, by considering them as systems of relationships, and by seeking the organizational principles that underlie all kind of systems (Biggart et al., 1998). According to Bogdanov, it is fundamentally wrong to separate out human activity from biological evolution and physical phenomena, and any such effort would result in significantly harsh consequences for our planet.

A second consequence of the self-organizing results from the importance of community whether it is the community of ants, bees or in a professional setting, a community of workers. The whole is not simply the sum of all the individual pieces and parts but something quite different.

This is the duty of our generation as we enter the 21st century – solidarity with the weak, the persecuted, the lonely, the sick, and those in despair. It is expressed by the desire to give a noble and humanizing meaning to a community in which all members will define themselves not by their own identity but by that of others.

Challenge Box: How do you react to the model of the universe as a community of interests rather than a collection of pieces and parts? How does that effect your understanding of the responsabilities of engineering?

4.9   TIME IN A MACHINE; TIME IN A COMMUNITY

Mechanical devices, assembly lines, natural laws of science and economics—they all point to the notion of the linear progression of time. Things proceed in an orderly, quantifiable way. We can identify initial conditions and boundary conditions and watch with delight as the process goes forward. In addition, we have a sense of being able to control the progression. We can twist knobs; fiddle with initial and boundary conditions to arrive at the optimum design. Take the mechanical clock for example. If we wish to slow it down, we can extend the length of the swinging pendulum or change the mass of the bob. On an assembly line, we can change the order of the parts being assembled or the workers who are employed. With the proper modifications, we can achieve whatever result we have decided to pursue. If we wish to build a resort city in the desert, say for example, Las Vegas, we can do just that. If we wish to prevent a river, for example the mighty Mississippi, from changing its path to the sea, we can do that as well for as long as we wish. With each technological advance, there is a sense that progress continues and with enough time we will be able to solve any and all problems that presently confront us. Such a view of the world seems very comforting, yet it is the product of a science which has long since been replaced.

Suppose an Objective Observer were to measure the success of Progress - that is to say, the capital-P myth that ever since the Enlightenment has nurtured and guided and presided over the happy marriage of science and capitalism that has produced modern industrial civilization.

Has it been, on the whole, better or worse for the human species? Other species? Has it brought humans more happiness than there was before? More justice? More equality? More efficiency? And if its ends have proven to be more benign than not, what of its means? At what price have its benefits been won? And are they sustainable?

With a shifting scientific paradigm, substituting the science of community for the science of mechanics, the physical property, time, remains the same but its significance for us in trying to understand natural processes and, in fact, our place in the universe is very different. Rather than speaking of a universal time, we would speak of an entire spectrum of time scales from incredibly small to the magnitudes associated with the size of the universe. In addition, rather than projecting time forward as a straight and true arrow, we would reflect upon the cyclical behavior of many phenomena and use the geometric shape, a circle, as a metaphor for that behavior.

Time depicted as a circle rather than an arrow decouples the passage of time from the myth of progress. With each new advance, we do not know where we are going ultimately and hence we cannot argue that the present will be necessarily better than the past or the present. It may or may not be depending in part on how we go about designing and implementing our proposed solutions.

Progress is the myth that assures us that full-speed-ahead is never wrong. Ecology, or the science of community, is the discipline that teaches us that it is a disaster.

Daniel Botkin, 1993

4.10   CONCLUSIONS

Buzz! The alarm clock just went off so we need to move on to the next thing scheduled and then the next and the next and so on *ad infinitum*! We explored the historical origin of our present conceptualization of time and discovered it is linked to a scientific paradigm that describes the universe as a mechanical device or machine. We also have discovered that science has changed and, of course, will continue to change. Nature now is most often conceived as a self-organizing system or a community of interests.

You, as engineers and technologists of the future, will be charged with developing solutions to new and ever more difficult problems. As future engineers, you can choose to model your world as a mechanical clock, treating all of nature including the animal and plant worlds as well as fellow human beings as collections of pieces and parts. More is always better, and the future can be controlled to meet our needs and satisfy our desires. Alternatively, you can choose to model your world as a community of interests, each with their own needs, shortcomings, hope and fears. The former world-view has led to inestimable wealth for a few and many unanswered questions concerning its sustainability in the long run. The latter approach represents a significant change in the ways we go about our professions. It suggests that more may be a recipe for disaster and we will never have certainty over the course of events, the impact of our engineering decisions. Above all else, it cautions for deliberateness and against arrogance.

Challenge Box: How does engineering change if you consider the importance of community in you work? Who is in your community? How far do the wall of your community extend?

CHAPTER 5

Driving the SUV

Figure 5.1. A green car?

"Human activities are increasingly altering the Earth's climate.... It is virtually certain that increasing atmospheric concentrations of carbon dioxide and other greenhouse gases will cause global surface climate to be warmer. The unprecedented increases in greenhouse gas concentrations, together with other human influences on climate over the past century and those anticipated for the future constitute a real basis for concern."
American Geophysical Union, 2003.

"Even though cars get worse gas mileage than two decades ago, they actually have become much more efficient. The problem is that the efficiency went into more power and larger, heavier vehicles, not into fuel economy. If all of the technological efficiency improvements had gone into efficiency, miles per gallon would be significantly higher than they are today."
Neal Elliott, American Council for an Energy Efficient Economy.

5.1   SO WHAT EXACTLY IS AN SUV?

SUV is the widely used and known acronym for 'sport utility vehicle.' At first conception, an SUV was a vehicle that combined the towing capacity of a full-size truck with the passenger and storage capacity of a minivan. However, as consumer demands have changed, so has the SUV. Many manufacturers now focus on fuel-efficiency and driving and riding comfort, rather than towing capacity.

© eVox Productions

Figure 5.2. An Example of an SUV (2008).

Typical features of an SUV include seating for five to seven, high seating and road positioning, roomy interior, non-dedicated trunk space, high engine capacity and 4-wheel drive capability. Though the SUV was originally designed to be an off-road vehicle for sporting purposes, their popularity has spawned several different breeds, including the luxury SUV.

As is the case for passenger cars, the SUV has different classes and sizes.

Though many modern day motorists value the SUV for its size and roominess, many others criticize their lack of fuel efficiency and their contribution to air pollution. Consumers who value the SUV do so not only for its size, but the perceived safety of driving such a bulky vehicle. Though crash test safety ratings vary with makes and models, some SUVs are known to pose the risk of rollover.

For those who enjoy the flexibility of combining weekday comfort with weekend fun, the SUV has proven to be a leading choice in vehicles. Similarly, having 4-wheel drive capability and towing capacity without the passenger restrictions of a truck appeals to many car buyers and keeps the SUV a leading seller.

As a concern for the environment and the demand for fuel-efficient vehicles increases, manufacturers worldwide are continuing to explore ways to make improvements to the SUV family to keep them a leading selling vehicle.

5.2   THE SUV AND THE ENVIRONMENT

SUVs represent a paradox to consumers - television advertisements present them as a way to return to nature, yet they actually accelerate existing environmental problems. Commercials often depict happy families driving on mountain roads, avoiding falling rocks and enjoying the flowered wilderness in leather-seated comfort. The sad truth is that these vehicles are contributing to the destruction of our natural resources. In reality, only 5 percent of SUVs are ever taken off-road, and the vast majority of these vehicles are used for everyday driving. In 1985, SUVs accounted for only 2 percent of new vehicle sales. SUVs now account for one in four new vehicles sold, and sales continued to climb until 2008.

Driving an SUV has a much greater impact on the environment than driving other passenger cars due to vastly different standards set by law and government regulations. For example, current federal regulations allow SUVs to have far worse fuel economy than other vehicles. The federal corporate average fuel economy (CAFE) standards set the fuel economy goals for new passenger cars at 27.5 miles per gallon (mpg). But under the law, SUVs are not considered cars - they are characterized as light trucks. Light trucks only have to achieve 20.7 mpg. It should be noted that this is an average for all light trucks, which is why it is possible to have SUVs on the road that only achieve 12 mpg. When CAFE was instituted in the 1970s, there were few SUVs and light trucks on the road, and they were primarily used for farm and commercial work. Today, however, the demographics of an SUV buyer are quite different. The amount of gasoline burned by a vehicle is important for several reasons. The most crucial is the threat of global warming.

5.3   THE THREAT FROM GLOBAL WARMING

Global warming has been extremely well studied. In 2001, the Intergovernmental Panel on Climate Change (IPCC) issued a report on global warming with many dire predictions. The World Meteorological Organization and the United Nations Environment Programme created the IPCC in 1988 to study the risks associated with global climate change. The IPCC found that about three quarters of the anthropogenic (caused by humans) emissions of carbon dioxide to the atmosphere during the past 20 years is due to fossil fuel burning. The IPCC anticipates higher temperatures and heat waves over the next century, as well as more intense and dangerous storms.

According to the EPA (Environmental Protection Agency), "increasing concentrations of greenhouse gases are likely to accelerate the rate of climate change. Scientists expect that the average global surface temperature may rise 1-4.5°F (0.6-2.5°C) in the next fifty years, and 2.2-10°F (1.4-

Figure 5.3. Contributing to global warming.

5.8°C) in the next century, with significant regional variation. Evaporation will increase as the climate warms, which will increase average global precipitation. Soil moisture is likely to decline in many regions, and intense rainstorms are likely to become more frequent. Sea level is likely to rise two feet along most of the U.S. coast."

The first reliable global measurements of temperature from NASA, published by Hansen and his colleagues in 1981, showed a modest warming from 1880 to 1980, with only a slight dip in temperatures from 1940 to 1970 [Graph adapted from Hansen et al. (1981)].

In 1981, NASA scientists predicted the impact of carbon dioxide emissions on global temperatures between 1950 and 2100 based on different scenarios for energy growth rates and energy source. If energy use stayed constant at 1980 levels (scenario 3, bottom lines), temperatures were predicted to rise just over 1°C. If energy use grew moderately (scenario 2, middle lines), warming would be 1–2.5 °C. Fast growth (scenario 1, top lines) would cause 3–4°C of warming. In each scenario, the warming was predicted to be less if some of the energy was supplied by non-fossil (re-

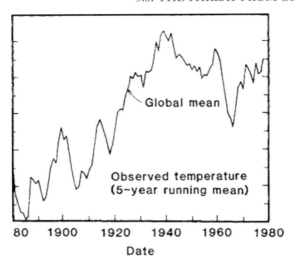

Figure 5.4. Historical record of global mean temperature.

newable) fuels instead of coal-based, synthetic fuels (synfuels) [Graph adapted from Hansen et al. (1981)].

According to the EPA, one of the most important things you can do to reduce global warming pollution is to buy a vehicle with higher fuel economy. Every gallon of gasoline your vehicle burns puts 20 pounds of carbon dioxide (CO_2) into the atmosphere. Scientific evidence strongly suggests that the rapid buildup of CO_2 and other greenhouse gases in the atmosphere is raising the earth's temperature and changing the earth's climate with potentially serious consequences. Choosing a vehicle that gets 25 rather than 20 miles per gallon will prevent 10 tons of CO_2 from being released over the lifetime of your vehicle. Passenger cars and trucks account for about 20 percent of all U.S. CO_2 emissions.

Note that a vehicle which gets 10 mpg emits approximately 120 tons of CO_2 while one that gets 30 mpg emits less than half that amount. The National Academy of Sciences estimates that if fuel economy had not been improved in the late 1970s, U.S. fuel consumption would be about 2.8 million barrels of oil per day higher than it is. Unfortunately fleet-wide improvements in vehicle fuel economy occurred from the middle 1970s through the late 1980s. Since then fuel economy has been consistently falling. In fact, average new vehicle fuel economy fell in 2000 to 24 mpg, its lowest level 20 years primarily due to the increase in numbers of SUVs.

For more information, the EPA has a website devoted to global warming, which answers many of the common questions about the problem. (http://www.epa.gov/globalwarming/).

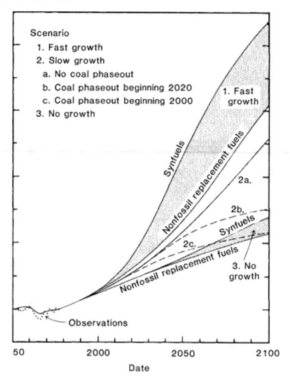

Figure 5.5. Growth of CO_2 for various scenarios.

5.4   SUV'S SMOG FORMING EMISSIONS

SUV's have a significant environmental impact even beyond the problem of global warming. Federal law gives heavy sport utility vehicles permission to emit higher levels of toxic and noxious pollution - carbon monoxide, hydrocarbons, and nitrogen oxides. Sport utility vehicles emit 30 percent more carbon monoxide and hydrocarbons and 75 percent more nitrogen oxides than passenger cars. These combustion pollutants contribute to eye and throat irritation, coughing, nausea, dizziness, fatigue, confusion and headaches. Hydrocarbons and nitrogen oxides are precursors to ground level ozone, which causes asthma and lung damage. Unfortunately, increasing numbers of Americans are living in areas with poor air quality from ozone pollution, according to the American Lung Association (ALA). The ALA found that 141 million Americans lived in areas with poor air quality during 1997-1999, an increase of nine million since 1997-1999.

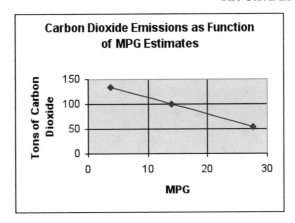

Figure 5.6. Tons of carbon dioxide emitted over vehicle lifetime.

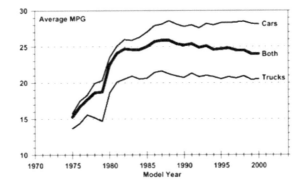

Figure 5.7. Fleet-wide Fuel Economy from 1975 to 2000.

5.5 U.S. DEPENDENCY ON OIL

Finally, it is important to note that SUVs are contributing to our dependence on imported oil. The more gasoline we use, the more oil we have to import from other countries. Currently, more than half of the oil we use is imported.

Figure 5.8 depicts the increase in the importation of foreign oil over the course of the last half century. Note that in 1993, the U.S. imported more oil than it produced and that trend has grown continuously.

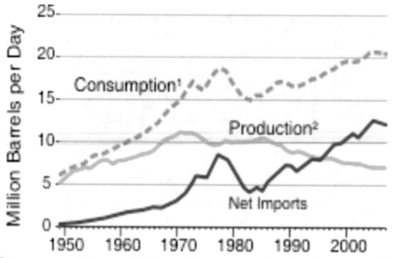

Petroleum products supplied is used as an approximation for consumption.
Crude oil and natural gas plant liquids production.
Source: Energy Information Administration. *Annual Energy Review 2007—*
Table 5.1. (June 2008)

Figure 5.8. Production, Consumption and Importation of Oil in the U.S.

5.6   WHAT CAN BE DONE?

In July 2001, the National Academy of Sciences (NAS) released a study on fuel economy standards. The NAS found that light trucks, SUVs, minivans, and pickup trucks could reach 28-30 mpg for an additional cost of $1,200-$1,300. The Union of Concerned Scientists (UCS) has also looked at this issue, and found similar results. The UCS report concluded similar fuel economy levels were achievable at nearly identical consumer cost - all using existing technologies that automakers could implement quickly.

Buying a sport utility vehicle can be tempting but it's important to understand that it may mean extra expenses. Automakers have been able to charge high prices for sport utility vehicles (SUVs) making them very profitable. At the beginning of 2008, the light trucks - including SUVs - now make about half of vehicle sales in the U.S. But light trucks provide about two-thirds of profits for the Big Three automakers. While U.S. automakers make little or no profit on other passenger cars, a large SUV can bring $12,000 to $20,000 in SUVs often have heavier maintenance costs, greater gasoline bills, and higher insurance rates than other cars. In addition, recent research shows that even minor accidents can result in tremendous repair costs. The reason that the SUVs cost so much to repair has to do with federal regulations. Passenger car bumpers have to meet federal standards in low-speed crashes, and most of the bumpers on passenger cars include a reinforced bumper bar and foam to absorb crash energy. But SUVs are not subject to any kind of bumper requirements, so they are allowed to crumble in low-speed accidents.

More generally, taxpayers are also bearing an increased burden from the growth in SUV sales. This is due to a loophole in the federal tax code that exempts SUVs from a tax imposed on other cars. Known as the "gas guzzler tax" - it was instituted in 1978 to encourage automakers to manufacture fuel-efficient vehicles. More efficient vehicles pay little or no tax. Less efficient vehicles pay more tax. Except for SUVs which pay no tax. Several environmental groups have estimated that this loophole costs approximately $1.1 billion in revenues that must be made up by the typical taxpayer. As a result, we all pay more taxes while automakers produce more polluting vehicles.

5.7   CASE STUDY

Scenario: Market analysis has shown that there is a rapidly increasing demand for SUVs in various third-world countries. Several large multi-national corporations have decided to first import and then locate manufacturing and assembly plants for the SUVs in the various countries. The SUV is seen as an important element in the economic development of the countries by the ruling governments. You are a design engineer for the largest corporation and are charged with developing an 'appropriate' design and prototype of the proposed vehicle.

Approach: Using the various decision making models described in Book 2, let us explore the kinds of questions that such models challenge us to include in our deliberations and ultimately our design.

- Engineering and Freedom

Consider first an engineering decision model based upon freedom. We considered three kinds of freedom: existential freedom, substantial freedom and the freedom of choices. According to Sartre, that is, freedom as constructed from an existentialist perspective, must take on the responsibility of choosing for all of humankind, desire and work for the freedom of all humanity, and create ourselves within the context of the relationships and obligations we have to others. In describing substantial freedom Sen states that "the perspective of freedom" is concerned with "enhancing the lives we lead and the freedoms we enjoy." The ethic of freedom calls for, "expanding the freedoms we have reason to value," so that our lives will be "richer and more unfettered" and we will be able to become "fuller social persons, exercising our own volitions and interacting with–and influencing–the world in which we live." Nussbaum describes, "a twofold intuition about human beings: namely, that all, just by being human, are of equal dignity and worth, no matter where they are situated in society, and that the primary source of this worth is a power of moral choice within them, a power that consists in the ability to plan a life in accordance with one's own evaluation of ends."

So what kinds of questions should we consider in our SUV design deliberations? Sartre might suggest that we consider if we would want more SUVs here on our roads. Can we accept more congestion, more pollution, more noise and more dependence on fossil fuels? Sen might challenge us to consider whether or not more SUVs in the various countries expand the freedoms of the native populations at all? Are the citizens more able to live their lives or are their lives more restrained or constrained by the presence of the new vehicles? Both might ask us to fully explore the notion of economic development. What does that term actually mean and who benefits from such development? Nussbaum might caution us to consider whether or not all members of society are considered to have "equal moral worth" and ask whether or not the SUV promotes or at a minimum does not prevent citizens from living a life consistent with their own evaluations.

- Engineering and Chaos

Our understanding of chaos led us to the following prescriptive principle for making engineering decisions:

> *A thing is right when it tends to allow the natural world and all the entities thereof, to thrive in richness and diversity, and to experience change. It is wrong when it tends otherwise.*

Engineering decisions using a chaos perspective caution us to focus on the health of the various ecosystems and the natural world in general. What are the impacts of the SUVs on the ecosystems were they are produced? What about where they are used? What are the short term impacts? What are the longer term impacts? What effects will the increase in SUVs have on the various cultures–the culture in which the SUV is produced and that culture in which more and more are put on the road?

- Engineering and a Morally Deep World

A decision model based on a morally deep world suggests a self-organizing system (our planet) is characterized by synthesis rather than analysis and suggests a new code of responsibility based upon community rather than individuality. For a morally deep world, the first fundamental canon and rule of practice is specified as:

> *Engineers, in the fulfillment of their professional duties, shall hold paramount the safety, health and welfare of the identified integral community.*

The fundamental difference between an ethical code based on a morally deep world versus our present sense of responsibility is the replacement of the "public" by the "identified integral community." Who then is included in the integral community?

  - Multinational corporations? Their shareholders? Their employees?
  - People who will drive the SUVs? Those that will service them?
  - The ecosystems that surround the various production facilities? The ecosystems that are affected by the shipping of the SUVs?
  - The indigenous communities in which they will be sold? Their customs and ways of life?
  - Anyone else?

- Engineering and Globalism

As we discussed earlier, globalization in its literal sense is the process of making, transformation of some things or phenomena into global ones. It can be described as a process by which the people of the world are unified into a single society and function together. This process is a combination of economic, technological, socio-cultural and political forces. The question that we are asked to consider is the following: Is the production and use of more SUVs consistent or in opposition to the *Global Ethic*? By *Global Ethic* is meant the necessary minimum of common values, standards and basic attitudes or alternatively a minimal basic consensus relating to binding values, irrevocable standards and moral attitudes.

• Engineering and Love

In order to understand the nature of an engineering based on love, we sought wisdom from Thomas Berry. Berry's most famous quotation is:

The Universe and thus the Earth is a communion of subjects, not a collection of objects.

By communion, Berry was referring to intimacy or a feeling of emotional closeness, a connection, especially one in which something is communicated or shared. By subject, the reference is to the essential nature or substance of something as distinguished from its attributes. According to Berry, our new community is a very special one; that is, it is one in which the various elements are bound together as subjects having interests rather than one in which some have interests while others are simply resources to be utilized. What are the implications then for our SUV project? Have we considered those who build, those who ship, those who buy, and those who will use our SUV as objects or as subjects? Are they simply cogs in our wheels of career advancement? Economic development? And what of the various ecosytems? How will they be impacted?

CHAPTER 6

Travelling to Waikiki Beach

Figure 6.1. Going to the beach with the kitchen sink.

The native language is soft and liquid and flexible and in every way efficient and satisfactory–till you get mad; then there you are; there isn't anything in it to swear with. Good judges all say it is the best Sunday language there is. But then all the other six days in the week it just hangs idle on your hands; it isn't any good for business and you can't work a telephone with it. Many a time the attention of the missionaries has been called to this defect, and they are always promising they are going to fix it; but no, they go fooling along and fooling along and nothing is done.

Mark Twain's Speeches, 1923 ed. "Welcome Home."

6.1   INTRODUCTION

Mention Waikiki Beach in Honolulu, Hawaii, and we immediately think of beautiful white beaches, breathtaking panoramas, waves that meet the needs of even the most adventuresome surfer, and throngs of tourists parading along the sea-side promenade, thousands more busily going from exclusive boutique to novelty store, arms loaded down with countless boxes and shopping bags. Gaze out to sea and there is the magnificent Diamond Head peak looming ominously above the morning mists. Everywhere one looks there are the beautiful people of wealth and high fashion. Certainly, Waikiki Beach is nothing less than paradise here on Earth or is it? What about the people who inhabited the island long before Captain Cook landed?

Figure 6.2. Diamond Head as seen from Waikiki Beach.

Technological advancements have played an important role in rapid economic development and huge growth of tourism on the islands we refer to as the State of Hawaii. We shall examine both the environmental as well as societal impacts. Rapid advances in technology have had profound impacts on the Hawaiian people. We shall attempt to examine those impacts in light of many of the ideas we discussed for making engineering decisions in the 21st century (Book 2).

6.2   HISTORY

Polynesians, ancestors of the Hawaiian people, undertook the long ocean voyage from the Marquesas Islands to Hawaii at least 1,700 years ago. At European contact in 1778, an estimated 400,000 to 800,000 Hawaiians lived in a society with highly complex political and social systems. Separate high chiefs governed the major islands, with subordinate chiefs managing self-sustaining land units encompassing broad plains near the sea, running up valley ridges to the mountains. Within these units, the people used the resources necessary to sustain life—access to offshore fishing and shoreline gathering; plots of land and sufficient water for growing taro, banana, breadfruit, or sweet potatoes; the right of way to the uplands for timber and fuel; and the right to hunt and gather wild plants and herbs.

This would all change as a result of contact with Europeans. On his third voyage into the Pacific, the explorer Captain James Cook, British commander of HMS Resolution and HMS Discovery, on January 18, 1778 found Oahu and Kauai. He was thought of by the Hawaiians as the reincarnation of Lono, one of their principal gods. Cook named Hawaii the Sandwich Islands in honor of the Earl of Sandwich. He returned to Hawaii a year later and was slain there on February 13, 1779. As a result of the European contact, contagious diseases including cholera, measles and gonorrhea, decimated the Hawaiian population. The population was estimated at between 250,000 to 1 million when Captain Cook sailed into Kealakakua in 1779. By 1848 Hawaiians numbered 88,000. By the year 2044, demographers predict that there will be no more Native Hawaiians left.

Figure 6.3. Captain James Cook.

6.3   BACKGROUND

Many Native Hawaiians share the view that tourism, as a foreigner dominated enterprise, is the plague which an already oppressed people must endure with very few other economic options or alternatives in life. Still, many end up choosing the lesser options even if it means unemployment or criminal activity. It is no accident that Hawaiians are the poorest of all people in Hawaii, with the highest percentage of unemployment, welfare recipients and prison populations. Notwithstanding the tourist industry, Native Hawaiians continue to be the poorest, sickest and least educated of all people in Hawaii.

Tourism is concerned with self-preservation as an industry and not with the well-being of the community. In March of 1991, during the dramatic decline of visitors to Hawaii due to the Gulf War, the Hawaii State Legislature readily allocated as an emergency measure $6 million to be used by the Hawaii Visitors Bureau for television commercials on the mainland USA. During the same period, hundred of hotel employees were laid off in one of the largest layoffs in recent years. No emergency measures to assist the unemployed were introduced or even considered.

Foreign investment related to tourism went from 70.8 million dollars in 1981 to over a billion and a half in 1986. Japanese investment is over 3 billion dollars for hotels alone from 1989-2007. The Australians remain in second place with 117 million invested. Today, almost every major hotel is owned by foreign investors and nearly every hotel planned is being funded by foreign investment.

Business interests caused the illegal overthrow of the Hawaii Nation in 1893. A hundred years later, the same business driven interests continue and Native Hawaiians remain victims of an exploitation whose guise is an industry called tourism. A basic human right is the ability of a people to be self-governing, self-determining and self-sufficient. This right was taken away from Hawaiians when the nation was overthrown. Tourism in many respects perpetuates the oppression.

In its current form, tourism has evolved to a point where it is of minimal economic consequence to Native Hawaiians. Tourism, which stifles economic diversification and weakens existing agricultural and technological development, does not provide a viable economic alternative to Native Hawaiians.

It is interesting to consider the increase in the number of hotel rooms since 1985 (Fig. 6.4). Along with this data, consider the increases observed in actual and projected visitor numbers (Fig. 6.5).

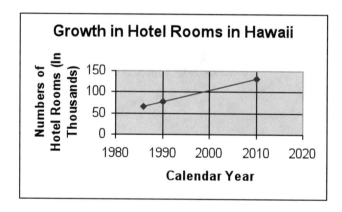

Figure 6.4. Growth in Numbers of Hotel Rooms in Hawaii since 1986.

Hawaii, it seems, is headed toward a non-diversified economic future that will be totally dependent upon the tourist industry. Tourism therefore, will continue to be a major obstacle in the movement of Hawaiians towards self-governance and self-determination. More poverty and the continuation of the negative impact upon Native Hawaiians will be the inevitable outcome of the future of tourism in Hawaii.

When the primary means of promotion is dependent upon a culture and people, and the perception that "all is well in paradise" is put forward while in fact "all is not well," then the issue becomes one of cultural prostitution. It becomes the selling of an artificial cultural image that has complete disregard for the truth, at the expense and pain of Native Hawaiians who are struggling

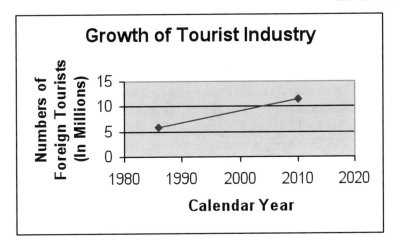

Figure 6.5. Increase of Foreign Visitors to Hawaii since 1986.

to survive. From printed brochure to life in the fast lane, tourism promotes the development and practice of an entertainment and visitor oriented culture. The follow-through with marketing and promotion is part and parcel of the "plastic tikis, Kodak hula, and concrete waterfalls."

Tourism development in Hawaii most often takes place at the expense of a people's cultural and historical symbols and land based resources. Tourism development has played a major role in the destruction of ancient Hawaiian burial grounds, significant archaeological historic sites and sacred places. Almost every major resort development has been built on some culturally significant site. Community opposition is usually based upon these cultural issues. The usually insensitive approach and manner of development leaves the local community to conclude that there is no respect or concern for the culture and identity of Hawaiian people.

The third major impact of tourism on Native Hawaiians must be understood in the context of environmental exploitation. The character of indigenous Pacific cultures in relationship to the land is one based on a high level of environmental awareness and ecological conservatism. The relationship of people to land, and people to sea, is spiritual and religious. Land is the base around which a culture evolves. When tourism takes away the land, takes away access to the fishing grounds or the right to gather food or medicine, the Hawaiian loses a primary means of livelihood, and more importantly, meaning in life. Crowded beaches and commercial tour boating do much to threaten shoreline or coastal fishing through noise or chemical pollution. The state has begun to identify beach parks and near-shore areas that are exceeding capacity use because of significant resident and visitor numbers. User conflicts between residents and visitors are becoming a problem and are expected to escalate as tourism and ocean recreation industries continue to grow.

There is a growing frustration among Native Hawaiians. This expresses itself in the very visible opposition to resort of related development and the increasing amount of land being openly occupied by indignant Hawaiians. Almost every large resort development in the last ten years has been opposed by Hawaiian groups or organizations.

The emerging awareness of the impact of tourism on Native Hawaiians captured local and international attention when, in August 1989, an international conference on tourism sponsored by local, national and international church groups and organizations took place in Hawaii. That conference looked at the negative impact of tourism on Native Hawaiians. The results were published in what came to be called the Hawaiian Declaration of 1989. The story of the Native Hawaiian people, a people who love their land, is a complicated and difficult one. But when told in broad strokes, it is a familiar one: a story of an indigenous people and of greed, racism, and imperialism.

6.4   CASE STUDY

Scenario: Market analysis has shown that there will be a huge increase in tourism and foreign visitors to Hawaii. This increase is in part due to the threat of terrorism in Europe and the Middle East. The airport at Honolulu is in need of significant improvements in its runways, which will require the acquisition of large tracts of land adjacent to the existing airport and several large parcels at other locations in the city. You are a design engineer employed by the multi-national engineering firm, which won the contract for the renovations and expansions. Your firm is world renowned for its strict adherence to principles of green design.

Approach: Using the various decision making models described in Book 2 and in the last case study, let us explore the kinds of questions that such models challenge us to include in our deliberations and ultimately our design.

- Engineering and Freedom

Consider first an engineering decision model based upon freedom. We considered three kinds of freedom: existential freedom, substantial freedom and the freedom of choices.

So what kinds of questions should we consider in our airport renovation and expansion design deliberations? Should we consider the impacts of increased tourism? Do we consider the effects on the environment? The native populations? Their history? Their culture?

- Engineering and Chaos

Engineering decisions using a chaos perspective caution us to focus on the health of the various ecosystems and the natural world in general. What are the impacts of the increase in tourism and foreign visitors on the local ecosystems? The health of the planet? What are the longer term impacts? What effects will the increase in foreign visitors have on the various cultures?

- Engineering and a Morally Deep World

The fundamental difference between an ethical code based on a morally deep world versus our present sense of responsibility is the replacement of the "public" by the "identified integral community." Who then is included in the integral community?

   o Multinational corporations? Their shareholders? Their employees?

   o Foreign visitors?

   o Native Hawaiians?

   o Local ecosystems? Endangered plants and animals?

   o Anyone else?

• Engineering and Globalism

  As we discussed earlier, globalisation, in its literal sense, is the process of making, transformation of some things or phenomena into global ones. The question that we are asked to consider is the following: Is the increase in tourism and foreign visitors consistent or in opposition to the *Global Ethic*? By *Global Ethic* is meant the necessary minimum of common values, standards and basic attitudes or alternatively a minimal basic consensus relating to binding values, irrevocable standards and moral attitudes.

• Engineering and Love

  In order to understand the nature of an engineering based on love, we sought wisdom from Thomas Berry. According to Berry, our new community is a very special one; that is, it is one in which the various elements are bound together as subjects having interests rather than one in which some have interests while others are simply resources to be utilized. What are the implications then for our airport renovation and expansion project? Have we considered all those who will be impacted as objects or as subjects? Are they simply cogs in our wheels of career advancement? Economic development? And what of the various ecosystems? How will they be impacted?

Bibliography

Abdala, A., pp. 285–310, Reunión Anual, 32, Bahia Blanca, 19-21 Noviembre 1997; Bahia Blanca, AAEP, 1997.

Aiello, R. and Grajales, F., Social aspects of Solid Waste management: The experience of Argentina, World bank Urban Development website, World Bank Group 2001.

Alcazar, Lorena; Abdala, Manuel A.; Shirley, Mary M., The Buenos Aires Water Concession, Volume 1 World Bank Report number WPS2311, 2000.

Amartya, Sen, Development as Freedom, Anchor Press, 2005.

Bak, Per, How Nature Works: The Science of Self-Organized Criticality, Copernicus Books, 1996.

Barlow, M., Bluegold: The fight to stop the Corporate theft of the Worlds water, Newpress, 2003.

Barlow, M., 2007.

Barlow, M., Blue covenant: The Global water crisis and the coming battle for the right to water, Newpress, 2008.

Berry, T., The Dream of the Earth, Sierra Club, October 2006.

Biggart, John, Glovelli,Georgii, and Yassour, Avraham, Alexander Bogdanov and the Origins of Systems Thinking in Russia. Avebury, 1998.

Bogdanov, A., The Struggle for Viability, Xlibris Corporation, 2002.

Botkin, Daniel, Discordant Harmonies, Oxford University Press: New York, 1993.

Bunge, Mario, Causality: The place of the causal principle in modern science, World Publishing Co., 1963.

CEAMSE, Estudio de calidad de los residuos solidos urbanos, CEAMSE, Enero 2007.

Cook, Steve, "More Americans Breathing Dirty Air, Lung Association Says in 2001 Report," Daily Report for Executives, Bureau of National Affairs. May 2, 2001. Page A-11.

http://www.deephawaii.com/hawaiianhistory.htm

Effectiveness and Impact of Corporate Average Fuel Economy Standards. Division on Engineering and Physical Sciences, Board on Energy and Environmental Systems, Transportation Resource Board. National Research Council, July 2001. Page ES-4.

Environmental Justice in Hawaii: *A Hawaiian Issue,* Rev. Kaleo Patterson, `http://members.tripod.com/~MPHAWAII/Tourism/EnvironmentalJusticeInHawaii.htm`

Francis, R., Water justice in South Africa: Natural Resources Policy at the intersection of Human Rights, Economics and Political Power, Berkely Electronic Press (`law.bepress/.com/expresson/eps/518`) 2005.

Fuel Economy website. Department of Energy and Environmental Protection Agency (`http://www.fueleconomy.gov/feg/climate.shtml`)

Galiani, S., Gertler, P., Schargrodsky, E., Water for life: The impact of the privatization of water services on child mortality, Journal of Political Economy, Vol. 113, pp. 83–120, February 2005.

Gobierno de la Ciudad de Buenos Aires Equipo Tecnico del Plan Estrategico, Indicatores, Part 2, Version Preliminar, marzo, p3. 1998.

Greene, Brian, The Elegant Universe: Superstrings, Hidden Dimensions, and the Quest for the Ultimate Theory, Norton, 1999.

Hansen, J., D. Johnson, A. Lacis, S. Lebedeff, P. Lee, D. Rind, and G. Russell, 1981: Climate impact of increasing atmospheric carbon dioxide. Science, 213, 957-966, doi:10.1126/science.213.4511.957.

Hawking, Stephen, A Brief History of Time, Bantam Press: New York, 1998.

Hounshell, David, Assembly Line, Johns Hopkins Univeersity Press: Baltimore, MA 1984.

"Hurricane Katrina," `http://en.wikipedia.org/wiki/Hurricane_Katrina`

Intergovernamental Panel on Climate Change (IPCC), 2001.

"Summary for Policymakers: A Report of Working Group I of the IPCC, 2001." (`http://www.ipcc.ch/`)

James, Jamie, The Music of the Spheres: Music, Science, and the Natural Order of the Universe, Copernicus Books, 1993.

Kanbur, R., Development disagreements and water privatization: bridging the divide, `www.people/cornell/pages/sk145/papers.htm`) 2007.

Kirkpatrick, C., Parker, D., and Zhang, Y.-F., State versus private sector provision of water services in Africa: an empirical analysis, Presentation Centre on Regulation and Competition, 3rd International conference, Pro-poor regulation and competition: Issues, Policies and Practices, Cape Town, South Africa, Sept 2004.

Light-Duty Automotive Technology and Fuel Economy Trends 1975 Through 2000. Executive Summary. Robert M. Heavenrich and Karl H. Hellman. Advanced Technology Division, Office of Transportation and Air Quality. U.S. Environmental Protection Agency. Air and Radiation EPA420-S-00-003
(http://www.epa.gov/otaq/fetrends.htm 06 Jan 2001).

Loftus, A. and McDonald, D., Lessons from Argentina: The Buenos Aires Water Concession, Municipal Services project Occasional Papers Series Number 2, 2001.

Marshall, Alfred, The Principles of Economics, Prometheus Books, 1997.

Mastny, Lisa, Ed., Vital Signs 2006, W.W. Norton & Co., 2005.

McDonald, D. and Rulters, G., Rethinking privatization: Towards a critical theoretical perspective, Public Services yearbook, Part 1, Chapter 1, 2005/6.

Motor Vehicle Facts and Figures, 1997. American Automobile Manufacturers Association. p. 84.

Mumford, Lewis, Technics and Civilization, Harvest/HBJ Books: New York, 1963.

Naughton, Keith, "The Unstoppable SUV," Newsweek, July 2, 2001.

Nussbaum, M.C., Sex and Social Justice, Oxford University Press, 1999. http://astro.temple.edu/~dialogue/geth.htm

Osborne and Van Loon, 1998.

Roberts, Paul, "Bad Sports," Harper's Magazine. April 2001.

Samson, M., Dumping on women: Gender and privatization of waste management, SAMWU (South African Municipal Workers Union and Municipal Services project), 2003.

Sim, S. and Van Loon, B., *Introducing Critical Theory*, Icon Books, Duxford, UK, p. 165, 2000.

Smith, Adam, The Wealth of Nations, Bantam Classics: New York, 2003.

Tennyson, The Works of Tennyson, Bibliolife, 2009.

The 1989 Hawai'i Declaration of the Hawai'i Ecumenical Coalition on Tourism Conference.

The Stanford Dictionary of Philosophy, 2009.
TOURISM IN HAWAI'I: Its Impact on Native Hawaiians And Its Challenge to the Churches, http://members.tripod.com/~MPHAWAII/Tourism/The1989.htm

The Worldwatch Institute, State of the World 2006, Worldwatch Institute, 2006.

Tolle, Eckhart, The Power of Now, New World Library: New York, 2004.

Tourisms Negative Impact on Native Hawaiians, Rev. Kaleo Patterson,
http://members.tripod.com/~MPHAWAII/Tourism/TourismsNegativeImpact.htm

UNESCO, International Tsunami Information Center, 2005,
http://ioc3.unesco.org/itic/

Watts, Stephen, The Peoples Tycoon: Henry Ford and the American Century, Knopf Press, 2005.

Wolfe, Maynard, Rube Goldberg Inventions, Simon and Schuster, 2000.

World Bank Development indicators, World Bank webpage, table 3.10 Urbanisation, 1999.

Printed in the United States
by Baker & Taylor Publisher Services